CORY DOCTOROW

THE REVERSE CENTAUR'S GUIDE TO LIFE AFTER AI

Cory Doctorow is the author, most recently, of *Enshittification*. He is a blogger, journalist, science fiction writer, and activist. For more than twenty years, he has worked with the Electronic Frontier Foundation to safeguard and further our human rights online. He was the coeditor of the weblog *Boing Boing* for nineteen years and now maintains a daily(ish) newsletter at Pluralistic.net. He has written more than thirty books, including nonfiction, fiction, graphic novels, and even a picture book. Born in Toronto, he now lives in Burbank, California.

ALSO BY CORY DOCTOROW

Little Brother

Little Brother

Homeland

Attack Surface

Martin Hench

Red Team Blues

The Bezzle

Picks and Shovels

Fiction

Down and Out in the Magic Kingdom

Eastern Standard Tribe

Someone Comes to Town, Someone Leaves Town

With a Little Help

Makers

For the Win

The Rapture of the Nerds

Pirate Cinema

Walkaway

The Lost Cause

A Place So Foreign and Eight More

Overclocked

The Great Big Beautiful Tomorrow

Radicalized

Graphic Novels

Cory Doctorow's Futuristic Tales of the Here and Now

In Real Life

Unauthorized Bread (forthcoming)

Enshittification (forthcoming)

Nonfiction

Enshittification

The Internet Con

Chokepoint Capitalism

How to Destroy Surveillance Capitalism

Information Doesn't Want to Be Free

Content

Context

Essential Blogging

The Complete Idiot's Guide to Publishing Science Fiction

THE REVERSE CENTAUR'S GUIDE TO LIFE AFTER AI

MCD

Farrar, Straus and Giroux • New York

THE REVERSE CENTAUR'S GUIDE TO LIFE AFTER AI

How to Think About Artificial Intelligence— Before It's Too Late

CORY DOCTOROW

MCD
Farrar, Straus and Giroux
120 Broadway, New York 10271

EU Representative: Macmillan Publishers Ireland Ltd, 1st Floor, The Liffey
Trust Centre, 117–126 Sheriff Street Upper, Dublin 1, D01 YC43

Art on title page and part openers by Oleksandr
Poliashenko / Shutterstock.com

Library of Congress Control Number: 2026934663
ISBN: 978-0-374-62156-8

Designed by Gretchen Achilles

Our books may be purchased in bulk for specialty retail/wholesale, literacy,
corporate/premium, educational, and subscription box use. Please contact
MacmillanSpecialMarkets@macmillan.com.

www.mcdbooks.com • www.fsgbooks.com
Follow us on social media at @mcdbooks and @fsgbooks

10 9 8 7 6 5 4 3 2 1

For the Electronic Frontier Foundation,
and everyone else who ever fought
for better technology

Your scientists were so preoccupied with whether they should, that they didn't stop to consider whether they could.

—MATTHEW W. BRADLEY

Contents

THE REVERSE CENTAUR'S GUIDE TO LIFE AFTER AI

Introduction

....

In May 2025, the internet briefly chose a new, unfortunate main character: Marco Buscaglia, a Chicago-based freelance writer behind a syndicated summer reading list. Though the list was credited to him, he wasn't its principal author. Rather, the list was written by a chatbot.

We know this is true for two reasons. First, because ten of the fifteen books on the list were nonexistent. They were AI "hallucinations" (a term that AI boosters use because it sounds sexier than "errors"). Second, because Buscaglia confessed to using a chatbot, in an interview with the veteran tech reporter Jason Koebler, who cofounded the journalist-owned news site 404 Media.

In Koebler's article, Buscaglia admits that he used a chatbot to source the list, and blames the inclusion of nonexistent books on the list on his own oversight: "I do use AI for background at times but always check out the material first. This time, I did not and I can't believe I missed it because it's so obvious."

Buscaglia comes across as utterly mortified, telling

Koebler, "No excuses. On me 100 percent and I'm completely embarrassed. It's a complete mistake on my part . . . This is just idiotic of me, really embarrassed."

For a few days, the internet had a lot of fun with poor Buscaglia, from the baffled authors whose nonexistent books were recommended on his list to the legion of people who are sick to the back teeth of having AI jammed into their eyeballs by overcapitalized tech companies.

But Koebler's analysis went beyond Buscaglia's contrition. In follow-up stories and on the 404 Media podcast, Koebler dove deep into the context of Buscaglia's unfortunate moment in the internet's baleful glare.

Start with the list itself. The summer reading list was many lists and guides in a sixty-four-page "Best of Summer" supplement that was distributed to multiple daily newspapers by King Features Syndicate, a division of Hearst Communications, a massive and established publishing conglomerate.

Buscaglia wrote (or "wrote") the majority of those lists. This is a remarkable feat! Koebler has experience writing lists like these, having begun his career as an intern at *Washington Monthly* magazine, where a list like this would be assigned to a team of three interns, overseen by an experienced newsroom journalist and backstopped by a large and skilled fact-checking department.

I've done this kind of work myself. I got my start as a journalist with assignments from *Wired* magazine that had me devoting an entire month to evaluating dozens of multitools, speaking to experts and manufacturers' representatives, in order to pick my top three and write a brief paragraph about each. These were fun assignments, but they were also serious and involved endless back-and-forth with the fact-checkers, even down to providing proof of how many different blades and attachments each tool had.

To be frank, there's *no way* that Buscaglia could have written all those lists and summer guides (including a guide to hammocks that features a quote from a nonexistent professor of leisure studies who is credited with writing a scholarly journal article on "hammock culture"—that paper also doesn't exist).

In 2025, any editor who assigns a freelancer to "write" the bulk of a sixty-four-page summer guide is either tacitly or explicitly commissioning that writer to confine their efforts to prompting a chatbot to barf up a bunch of plausible verbiage and hammer it into shape.

Buscaglia was asked to do the work of dozens of writers, researchers, editors, and fact-checkers, on a short timescale that guaranteed that the resulting product would be riddled with AI-generated errors. We don't know how much Buscaglia was paid for this job, but it's a sure bet that it doesn't add up to the total salary of all the skilled professionals whose jobs he was asked to do.

In other words, Buscaglia was set up to fail. His job wasn't merely to oversee a chatbot—it was to absorb the blame for that chatbot's mistakes.

Like Buscaglia, I am a freelance writer. What's more, I'm a freelance writer who has recently used AI in the course of my work. I was trying to locate a quote from an expert I'd heard say something very smart on a podcast. The only problem was that I couldn't remember the name of the expert *or* which podcast I'd heard him on.

So I downloaded an open-source AI speech transcription model called Whisper and installed it on my laptop. Then I downloaded thirty or forty hours' worth of podcasts that I'd recently listened to, and fed them all to Whisper, asking it to produce a transcript of all of them.

While Whisper chugged away in the background on my laptop, I kept working on the article, answering emails and alt-tabbing to social media in case someone had uploaded a video of a guy putting a lemon up his nose (normal writer stuff, in other words).

A couple hours later, I opened the folder where Whisper dumps its transcripts and searched them for some phrases I remembered from the quote I was looking for, and located the file in minutes (it was Hagen Blix, coauthor of *Why We Fear AI*, discussing the plight of AI therapists on the excellent *This Machine Kills* podcast). Then I used the time code Whisper provided to open the podcast file and clean up the transcribed quote, which I pasted into my article (I was right; it was a great quote).

This was a *great* experience. It was a great *AI* experience. Thanks to AI, my computer has now acquired a permanent new feature: it can turn arbitrary amounts of recorded speech into pretty reliable transcripts, in a manner so efficient that my laptop's fan doesn't even turn on. The fact that Whisper is open-source means that it can be maintained forever, by anyone in the world who has the skills and desire to do so, no matter what happens to Whisper's manufacturer (a grossly overhyped and terrible firm called OpenAI).

Two freelance writers, using AI. One was made miserable and embarrassed by his AI usage. The other had a delightful experience, saved a bunch of time, and produced a better piece of writing. How to explain the paradox?

One possibility is that the difference lies in *how* we used AI. I used AI to transcribe some audio, whereas Buscaglia used it to generate some writing.

Though I don't brainstorm with AI or other forms of automated or random text generation, I don't have a moral objection to these practices. There are plenty of writers who find text generation to be a productive tool for producing excellent work. The Dadaists' "cut-ups" involved slicing phrases out of a manuscript and scrambling them around, looking for serendipitous juxtapositions. Writers delight in Brian Eno and Peter Schmidt's "Oblique Strategies" deck, which contains fifty-five cards printed with mysterious orders ("Be the first person to not do something that no one else has ever not done before"). The Surrealists loved their "Exquisite Corpse," in which a group of writers take it in turn to append text to the previous writer's text, but each writer can see only the last sentence of the work when they begin.

Today, I know lots of writers who use chatbots to produce fine work: they might use the chatbot as a sounding board for evaluating ideas (or variations on ideas), or to challenge them with writing prompts, or to suggest improvements. As I said, I don't use AI that way, but I know people who do, and I like the things they write.

The difference between Buscaglia's AI experience and my own—the difference between a useful tool and a technological torment—is the difference between a *centaur* and a *reverse centaur*.

In automation theory (the academic study of automation), a *centaur* is a person who is assisted *by* a machine. Think of a clever human's head, arms, and hands atop a horse's strong body. Riding a bicycle or driving a car makes you a centaur; so does using the mute button on your TV remote when an ad comes on. Wearing a hearing aid makes you a centaur, and so does using a calculator to multiply large numbers. Being a centaur can be glorious. My whole writing career has been a serious of centaur moves, from the used IBM Selectric my parents bought for me to play with when I was six or seven to the Apple II Plus we got when I was nine, which kicked off an unbroken string of better and better writing tools, with spellcheckers, version control, collaboration features, autosave, and more. Best of all, I get to choose *exactly* which of these features I use. I work for myself, after all, so if I don't want to use the grammar checker (hell no), that's my business. No one expects me to write more pages when I get a new tool. A couple of years ago, a former student of mine asked me to try out his LLM tool that would help me write dialogue and flesh out characters. I played with it for ten minutes and then never went back to it. No one told me I was being uncooperative or spoiling a grand plan to increase efficiency and realize cost savings by refusing to use AI. Sometimes, when I'm really stuck, I write with a pen in a notebook. No one cares, except me.

A *reverse centaur* is a human who is conscripted into acting as an assistant *to* a machine. There's a classic *I Love Lucy* episode where Lucy and Ethel are working on an assembly line at a chocolate factory, taking bonbons off the belt and wrapping them in paper. As the belt goes faster and faster, Lucy and Ethel have to work at superhuman speed. They're reverse centaurs: the machine can move the chocolates from one place to another, but it needs a human to pack them into the box, and the humans who act as its assistant are made to work at a pace that exceeds all human capacity until calamity ensues.

I Love Lucy played that bit for laughs. Amazon warehouse workers get the horror-movie version. They are observed by AI-equipped cameras that time their movements and monitor their "time off-task," penalizing them if they fail to make quota. Amazon warehouse workers experience the highest level of on-the-job injuries in the

U.S. warehouse sector, and many of them have to resort to urinating in bottles because a visit to the toilet would blow their quotas. An Amazon warehouse is full of machines, but there are jobs the machines can't do, and that's where the Amazon warehouse workers come in—they assist the machines. They are reverse centaurs, and they have been conscripted to serve as peripherals for the warehouse's automation systems. They aren't merely used by those machines—they are used *up*.

It's not hard to imagine how a warehouse worker might *choose* to use AI in their daily work—for example, a computer vision system in a pair of smart glasses highlights an item they're looking for in a bin (there've been many instances where I've stared directly at something without seeing it in which *I* would have loved this). Automation isn't *necessarily* the enemy of warehouse work: there's nothing wrong with a forklift! The difference between automation that helps a warehouse worker and automation that torments that worker is whether the worker gets to choose where, when, and how to use that automation. It's the difference between a centaur and a reverse centaur.

I love the automation system in my car that warns me if I'm drifting out of my lane, and I recently discovered (the hard way!) that if the person ahead of me brakes suddenly, my car will let out a sphincter-puckering series of beeps and activate its own brake. I was pretty happy about that, even if it did come as a hell of a surprise.

Compare that with the drivers in those Amazon vans

rolling around your neighborhood. They have to sign into at least *nine* separate apps, and they are continuously scored based on their driving performance, as assessed by various AI tools. Drivers lose points for braking or swerving (even if that's the only way to avoid a sudden road hazard) or for deviating from the proscribed route set by the AI (even if there are obstructions or hazards). Drivers have impossible-to-meet quotas and the per-parcel compensation rate drops if they fail to meet it. Drivers are forbidden from peeing in bottles, but also given no time to urinate. The driver is just a peripheral for the van, present only because the van can't drive itself or get your parcel onto your porch. They are a reverse centaur.

This centaur/reverse centaur distinction is the heart of the paradox at the heart of the debate about the usefulness of AI tools. When you find yourself surrounded by people swearing that a given tool is worse than useless and others swearing that it has made their lives easier and better, you can bet that the former group is made up of reverse centaurs who've had AI imposed upon them, the latter group is all centaurs who've gotten to make up their own minds about where, when, and how to use AI tools. The solution to the paradox is to stop thinking about what the gadget does, and pay attention to who the gadget does it *to* and who the gadget does it *for*. The important part isn't the technical characteristics of the device, it's the power relationships of the people who use the device.

It's been a long time since I last held a factory job.

I'm a science fiction writer by trade, which means that I get asked about AI about a million times per day. Some people think science fiction writers are seers or prophets (regrettably, some extremely delusional science fiction writers share this view). They imagine that science fiction is a literature that predicts the future.

This is nonsense, of course. The future isn't predictable, which is a damned good thing, because if the future were predictable, it would be foreordained, which would mean that the actions taken by people don't matter.

But science fiction writers *do* have an intimate relationship with the future. By creating futuristic parables—made-up stories about futures that may never come to pass—we SF writers remind everyone that the future is up for grabs, full of possibility.

That's in stark contrast to tech bosses, who claim total authority over our possibilities, both present and future. There's Mark Zuckerberg: *Yes, well,* obviously *you'd like to talk to your friends without my spying on you all from asshole to appetite, but let's be realistic here. That's like asking for water that isn't wet.*

Or Apple CEO Tim Cook: *Yes, certainly, it would be nice if we could provide you with a reliable mobile device without locking you into our App Store, where we rake in a 30 percent commission on everything you spend, and where we get to decide which apps you can use, and which ones you can't (like the privacy apps that we banned in China or the ICE-tracking app we banned in America). But come* on, *it's just not realistic to want a computer that a) works and b) takes orders from you, not me.*

This is a cheap bully's trick: insisting that their abusive behavior is out of their hands, that they are merely acting in accord with some kind of iron law or great force of history. Margaret Thatcher was the world champion of this: she insisted that her cruel program of austerity and privation was inevitable, repeating the phrase "There is no alternative" so often that wags started referring to her as "Tina" ("There *Is No* Alternative").

Thatcher's contemporary successors are to be found in the C-suites of the tech industry.

Call this philosophy *inevitabilism*, the insistence that there is only one conceivable way to do things, and any problems you're experiencing aren't anyone's fault, they're just inescapable *reality*. The fact that the world is filled with sneaky Bluetooth sniffers that track your location as you walk through shopping centers or even just down the street isn't something you can blame a person or a company for; it's just a fact.

Science fiction is an anti-inevitabilist literature. Science fiction is only incidentally about thinking up new gadgets and explaining what they do. Far more important than what the gadget *does* is who it does it *to* and who it does it *for*. As the legendary science fiction editor Gardner Dozois once wrote, "Most SF can predict the car, some SF can predict the drive-in theater, but SF that can predict the changes in teen-age sexual behavior as a result of the drive-in is vanishingly rare." To this I would add, "Truly visionary science fiction might think about how a world where you need a government-issued ID like a driver's license in order to engage in sexual experimentation might plausibly lead to a 'database nation' of ubiquitous surveillance."

The mere existence of a literature in which many different writers have imagined many different futures, each of which can *feel* possible and even desirable, is a rebuke to inevitabilism. The fact that social relations between people and their technology can be different means that the current arrangement is a choice, and—crucially—it means that we can choose something else.

AI hucksters want you to believe that all the things they call "AI"—an incoherent grab bag of many technologies, some of them not especially related to the rest—are coming, and that when they arrive, there is only one conceivable way that we could use them. This is just high-tech Thatcherism, the inevitabilist move of a bully who insists that they're only doing what implacable reality demands of them.

This is a book about what AI can and can*not* do, but even more important, it's about the possible social arrangements of AI, from not using some AI technology at all, to using it in ways that let some of us choose to be centaurs, while saving our friends and neighbors from being conscripted into reverse-centaurity.

The current wave of AI is full of software performing

impressive feats in both parsing and generating language and images and sounds. There are lots of interesting, fun, and productive ways to use this technology. There is nothing about the *technology* of AI that determines how it *must* be used. We can choose to use it sometimes, or never, or all the time, depending on our needs and proclivities. We don't have to let billionaires tell us how it must be used.

Margaret Thatcher's "There is no alternative" has a fine rejoinder in the science fiction writer William Gibson's famous maxim, "The street finds its own use for things."

SF writers make lousy prophets, but we can be pretty good technology critics.

BEFORE

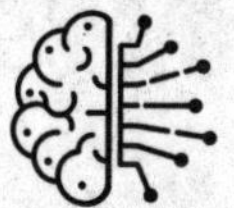

How We Got Here

. . . .

The AI hype machine *runs* on badly considered criticism, and it's not unique in this regard.

Lee Vinsel, a Science, Technology and Society scholar at Virginia Tech, described this process in a widely cited essay entitled "You're Doing It Wrong: Notes on Criticism and Technology Hype," which defined a critical misstep that Vinsel dubbed *criti-hype*, which he defines as "criticism that both feeds and feeds on hype."

When a tech company tells a fanciful story about the terrible things its products can do, they do so in anticipation of some advantage from convincing the public of their fearsome capabilities. For example, the advertising industry has long touted its ability to bypass our rational faculties in order to sell us things, aided by some kind of mind-control technique.

In the 1950s, ad companies provoked a moral panic by claiming that "subliminal advertising" could implant ideas directly into consumers' minds. They claimed that subtle images and words hidden in the shadows of illustrations, eyeblink single-frame insertions in films, and backward audio or audio that was too high or too low to be perceived by the human ear would act directly upon our desires, causing us to rush out and buy whatever they were selling.

It's easy to see why the advertising industry would want to tell this story: because it helps them sell ads. Just like your boss is primed to believe a story about how someday he'll be able to fire you and replace you with a docile chatbot, so, too, are merchandisers primed to believe that there is a tool that will cause you to march into their store, wallet in hand, and buy whatever they're offering at whatever price they're charging.

Many of the people who promoted the story of subliminal advertising doubtless believed it. Others may have been cynics who gave the clients what they wanted to hear.

Today, many of the practitioners of surveillance advertising techniques also doubtless believe their own stories (and Rasputin may have believed that he could hypnotize people with his mystical stare).

That doesn't make it true. The evidence for mind control via targeted advertising is awfully thin,* and Big Tech critics who breathlessly repeat these claims are unwitting accomplices to the sales departments of ad-tech

* See my 2020 book *How to Destroy Surveillance Capitalism* for more.

companies: *Why should you pay a 40 percent premium to advertise on Facebook? Just ask my critics: I've invented a mind-control ray.*

Contrast this with an anti-criti-hype approach: *Ad tech claims it has a mind-control ray, and it's using that outlandish claim to bilk advertisers out of hundreds of billions while amassing galactic-scale surveillance dossiers on every living person, which it both leaks and hands out to any cop or authoritarian state that asks for it. What they're calling "mind control" is, at best, just fine-grained targeting. They're not convincing people they're thirsty and then selling them a drink; they're listening in on billions of online conversations to find the people complaining about how parched they are and then showing them an ad for Coke.*

This is an approach that strikes directly at the source of ad tech's power, which comes from its profits, which come in turn from its ability to convince rubes that there's such a thing as a Big Data mind-control ray.

Moreover, this is an approach that (correctly) places the ad-tech industry on one side of a struggle, with advertisers (who are being sold a defective product) and advertisees (that's us, being spied upon and exploited) on the other side. Obviously, there's a certain irreducible degree of enmity between advertisers and the public, but we all share a common enemy in the ad-tech industry.

There's a certain kind of policymaker or governmental

enforcer whose priority is keeping businesses from being bilked by ad-tech hucksters. There's a different group of powerful people who are charged with protecting the public from surveillance. Framing ad tech as a scam—rather than as a miracle of persuasive technology—puts these two groups on the same side as well. It's a recipe for victory.

The companies that made billions selling (and mis-selling) ad tech are now on the vanguard of the AI bubble. It's the same scammers, pulling the same scam. The last time around, a bunch of tech's loudest critics got sucked into criti-hype, breathlessly warning about how our "dopamine loops" were being hacked by these evil sorcerers. Far from puncturing the ad-tech bubble, this criti-hype inflated it even further, making billions for the people currently driving AI psychosis among the world's deepest-pocketed, most easily gulled investors and policymakers. It will be downright embarrassing (not to mention terribly destructive) if we let these guys trick us into helping them sell their snake oil in back-to-back scams.

To win the AI fight, we need to enlist allies. If you're a worker whose job is in some AI pitchman's crosshairs, then every time you repeat claims about AI's current—or future—ability to do your job, you help that hustler convince your boss to fire you and replace you with an AI.

But it's worse than that, because repeating the AI barker's patter *also* alienates the people who benefit from your work. Some of those people will be grateful to you for the espresso shots you've pulled for them, or the contracts you've drafted for them, or the education you've provided to their kids, but a sizable fraction of this cohort will shed no tears for your technological obsolescence, not if it means that they're going to be able to get an unlimited supply of whatever you produce at prices so low they don't even register.

Here's an example that makes this distinction super clear: for decades, the largest, most profitable companies in the world have been outsourcing their customer service to overseas call centers. This move coincided with a trend toward steeply declining quality and thus a concomitant increase in how often you need to talk to a customer service rep. To make things even worse, these outsourced customer service reps don't work for the company and have severely curtailed agency and flexibility, meaning they generally can't solve your problem.

Long before AI came along, these people were being used by giant corporations as catchall accountability sinks. Their job was to get yelled at by you, at minimal expense to the company, which is why they earn terrible money and also why they can't give you a refund, change your plane ticket, or compensate you for the fact that your hotel room was covered in meth and dog hair.

No matter how astute you are, it's hard to see these people as class allies in a war against the company that indirectly employs them and directly shafts you.

But that becomes a lot easier when these human reps

are replaced by chatbots, who are so much more obviously put in place to simply absorb your frustrations and abuse without helping you.

In 2022, a Canadian named Jake Moffatt contacted Air Canada customer service to find out about bereavement fares so that he could travel to his grandmother's funeral. This was shortly after Air Canada had wired up some chatbots to its customer service desk, and one of these bots helpfully advised Moffatt that all he needed to do to make use of a bereavement ticket was to buy a regular, full-fare ticket, attend the funeral, and then submit a letter with proof of his grandmother's death, whereupon Air Canada would provide him with a refund for the difference between the full-fare ticket and the discounted bereavement fare.

Which is exactly what he did, only to be told by a (human) Air Canada customer service rep that the chatbot had "hallucinated" its advice about bereavement fares, and that company policy required fliers to provide proof of death *before* buying their tickets, and so no refund would be forthcoming.

Moffatt pursued this through every level of Air Canada customer service support, laboriously escalating his claim until he had exhausted all appeals. Finally, he filed a case with the British Columbia Civil Resolution Tribunal, and more than two years later, he was refunded CAD812.02.

There's never just one ant. The likelihood that the only lie an Air Canada chatbot told to one of its customers was this one is vanishingly small. Far more likely is that most of the Air Canada customers who lost money because one of the company's chatbots lied to them were simply unwilling to spend two years playing Air Canada's Kafka LARP just to get their money back. Their loss is Air Canada's gain.

Now, I happen to be Canadian (all the best Americans are, you know), and that means that I have spent hundreds of hours on hold with Air Canada, only to be connected to overseas call-center employees who were systematically curtailed from solving my problems. I am the *last* person to defend the old system of resolving Air Canada complaints.

But.

Those customer service reps—low-paid and ineffectual as they are—have repeatedly and sincerely tried to help me. When they made mistakes, it was due to misunderstandings. No one ever told me that the errors these people made were between me and them, and had nothing to do with Air Canada.

Someone sold Air Canada a chatbot by promising to put these people out of work. That same person knew—or should have known—that the chatbot would "hallucinate" a stream of incorrect advice that would cost Air Canada fliers like me money. In the war against Air Canada's remorseless enshittification, I am on the same side as those poor call-center folks.

Against Inevitabilism

· · · ·

The social arrangements of technology are a choice, not an inevitability. Consider the various iterations of the cash register: prior to the cash register, a grocery store clerk did skilled labor that made them hard to replace. While basic arithmetic isn't that hard to master, reliably summing up orders consisting of many items and making change for eight hours straight, without making major mistakes, requires a rare combination of talent and temperament.

From the grocery clerk's skill came power. The harder it is to replace a grocery clerk, the more the clerk can demand of their boss, including higher wages and better working conditions.

After a grocery store invests in a cash register, the clerk becomes much easier to replace, because the pool of people who can reliably punch buttons on an automatic tabulator and make change according to its calculations is much larger than the pool of people who can manage the same trick with pencil and paper.

This means that the boss can now lower the clerk's wages, add more duties to their workday, or fire half their clerks and give their work to the remaining half.

This isn't inevitable! The boss's freedom to fire a worker or change their job description or wages does not exist in a vacuum. It is socially determined: if the workers are in a union (or even if they aren't but they live in a country where the union movement has successfully lobbied for strong labor protections), the benefits of the cash register might be more equitably divided between labor and capital.

Depending on these social arrangements, a grocery clerk with access to a cash register might be freed from boring labor and given more time to chat with their customers. If the store attracts more customers, the clerk can serve those customers thanks to the labor-saving properties of the cash register, and demand a share of the higher profits that come from serving more customers without hiring extra staff. The boss who tells you that the only way to use a cash register is to fire your coworkers and make you do their jobs, too, is practicing vulgar Thatcherism. They're trying to bamboozle you with inevitabilism.

After all, there are many users of the cash register for whom it is an unalloyed good, like the grocery across the street that's organized as a worker's co-op, where the worker/owners enjoy the register for its labor-saving properties without having to fight with a boss over their share of the productivity gains of the new technology.

Successor technologies to the cash register—like mobile phone attachments that let people process credit card transactions—allow creative workers to sell directly to passersby at art fairs, comic-cons, and flea markets. They enhance the welfare and improve the material circumstances of workers. The most important thing about the gadget isn't what it does, it's who it does it *for* and who it does it *to*.

Inevitabilists will tell you a version of this story whose moral is, *See, you take the good with the bad*. They're wrong. You don't have to take the good with the bad. You can get the good *without* the bad. The difference isn't to be found in what a technology does; it's in what your boss uses the technology to inflict upon you. If you want to get the good without the bad, you need to switch from fighting *technologies* to fighting *bosses*.

Really Weird Math

. . . .

Your boss almost certainly loves AI. Bosses will not shut up about AI. Adopting AI, integrating AI, using AI—these subjects seem to weigh more heavily on our bosses' minds than objectively more important issues, like their quarterly profits. When bosses get so excited about a new tech trend that they forget about profits, it's a good bet that you're living through a tech bubble.

Even by the standards of tech bubbles, the AI bubble is a *whopper*. Hundreds of billions of actual dollars have been spent on AI hardware and data centers since OpenAI shipped ChatGPT 3 in 2020.

But even more hundreds of billions of *imaginary dollars* have been spent on AI. For example, Microsoft has a long-standing partnership with OpenAI. As part of this partnership, Microsoft gives OpenAI "tokens" that OpenAI can spend to access the computers in Microsoft's data centers.

OpenAI books these tokens as investment revenue, at face value. This is some very funny accounting. The tokens

Microsoft invests in OpenAI can only be redeemed for Microsoft data-center access. When Microsoft "invests" $10 billion worth of tokens in OpenAI, that "$10 billion" figure assumes that if you or I showed up at Microsoft and bought $1 worth of computing, and then Sam Altman showed up and ordered $10 billion worth of computing, Microsoft would simply multiply the amount of computing it sold to us by ten billion.

Think of it this way: say an ice cream cone costs $1 and contains one cup of ice cream. If you order ten billion cups of ice cream, that isn't $10 billion worth of ice cream. Anyone who orders ten billion cups of ice cream would expect a massive discount relative to the customer who buys just *one* cup—though of course it remains highly unclear what you're planning to do with all that ice cream.

So OpenAI is booking Microsoft's tokens as an investment that is grossly inflated—but that's just the warm-up.

The really weird math comes after OpenAI redeems its tokens with Microsoft to power ChatGPT and its other products: Microsoft books that transaction as $10 billion worth of AI-related *revenue* to its cloud-computing division.

Imagine that the ice cream parlor has some video-game machines on the back wall, and to play these, you need tokens that you can get from the cashier in exchange for dollars. But the cashier likes you, so they give you $10 worth of video-game tokens to play, which you promptly pump into the Galaga machine.

Using Microsoft and OpenAI's funny accounting, the ice cream parlor has "invested $10" in you (by giving you $10 worth of funny-money tokens), and then the ice cream parlor "earned $10 in revenue" (when you pumped the tokens into the machine).

Microsoft and OpenAI aren't alone in using these accounting gimmicks to create the impression that more investment and revenue are flowing into AI companies. Every AI company is using all kinds of cheap tricks to juke the stats to make AI seem like a bigger phenomenon than it actually is.

For example, you will often see established tech companies like Meta and Google trumpeting the amount of "engagement" their AI products received during the previous quarter. A naive person—or a credulous investor— might reasonably assume that this means that AI is popular among users.

But that is far from the truth. In a paper published in June 2025 in *The Journal of Hospitality Marketing & Management* with the catchy title "Adverse impacts of revealing the presence of 'Artificial Intelligence (AI)' technology in product and service descriptions on purchase intentions: the mediating role of emotional trust and the moderating role of perceived risk," two business-school professors reported on their work surveying everyday people about their attitudes toward AI. They concluded that more than 90 percent of us are *less* likely to use a product that's advertised as AI-based or AI-enabled.

If people don't like AI, what accounts for those glorious quarterly numbers about "increasing AI adoption" and "gains in AI interactions"?

As with centaurs and reverse centaurs, the seeming paradox disappears once you apply the lens of incentive and power to the system. To unpack those power relationships and incentives, we need to step out of AI for a moment and expand our focus to look at the tech industry that gave us the AI bubble.

Where do AI products come from?

Tech companies are among the most *instrumented* firms in world history. Tech companies pioneered the use of "productivity monitoring" tools for their tech staff, counting keystrokes per hour or lines of code per day.

Savvy tech bosses have long known that these measurements were at best imperfect and even counterproductive. You don't want your coders to type as many lines as possible—you want them to type exactly the right number of *correct* lines. Famously, a young Bill Gates mocked IBM's practice of awarding bonuses to programmers based on the number of lines they generated as "the race to build the world's heaviest airplane."

But while many tech bosses moved past measuring productivity by counting keystrokes, they continued to market those counterproductive productivity tools to their customers. Bill Gates may sneer at "the race to build the world's heaviest airplane," but a key selling point for

Office365, Microsoft's flagship enterprise cloud productivity suite, is a set of monitoring tools that provides bosses with a dashboard that ranks their juniors' activities—how often they type, click, and backspace; how many times they open, close, or create a new document.

Managers can find out how individual employees stack up against one another, see how their departments compare, and even compare their company's "productivity" with productivity at rival firms in their sector (apparently, no one who uses this feature to see proprietary information about their direct competitors ever worries that this means that those direct competitors can see *their* data).

Meanwhile, tech workers may have largely escaped the kind of granular monitoring that their companies inflict on less-technical workers, but their bosses haven't given up on measuring their performance. After all, the reason that tech *workers* are so easy to monitor is that tech *products* are so good at monitoring their users.

The digital tools you use can gather endless statistics on your usage. The Instagram app, for example, tracks everything from how quickly you scroll, what's on the screen when you stop scrolling (or just slow down), even readings from your phone's accelerometer that tell Meta how you're holding and moving your phone while you interact with their app. All this can be correlated with your location, the specifications of your device, and your other activities—the places you've visited recently, the places you subsequently visit, the things you buy, and the people you converse with.*

The fact that your technology tools spy on *you* means that the tech workers who *make* those tools can be rewarded or punished based on your usage.

Perhaps you're old enough to remember Google+, which was Google's 2011 full-court-press effort to launch a social media service to compete with Facebook. Google's product teams were ordered to find ways to integrate G+ into their products (for example, by shifting YouTube comments to Google+ message boards). To impress a sense of urgency upon their employees, Google management made Googlers' bonuses and stock options contingent on how often the users of their products interacted with Google+ (it's common for techies' bonuses to account for much, even most, of their annual pay).

In tech management jargon, G+ interactions were made into the company's overriding "Key Performance Indicator," the number that everyone had to hit in order to end the year in the black. KPIs are used in many sectors, but tech is uniquely amenable to them because

---

* Note that none of this is inevitable. Meta could absolutely let you look at Instagram without any of this surveillance. In 2022, two teenagers built their own Instagram app called "OG App" that did just that. OG App reached the top ten download list on both the Google and Apple app stores within a day, and then Meta, Google, and Apple colluded to shut it down. What your technology does is less important than who it does it for and who it does it to.

everything a user does with a product can be monitored and recorded, creating many categories of usage statistics that can be turned into KPIs.

KPIs definitely have a focusing effect on the working lives of tech workers, but they are among the most perverse of all incentives. A piece of common wisdom goes, "You treasure what you measure." Or, more pointedly, there's the version known as Goodhart's Law: "When a measure becomes a target, it ceases to be a good measure."

If we're being generous, we can imagine that the Google bosses who turned G+ interactions into a KPI were thinking, "If users are interacting with G+ more, that means that they're enjoying it." Even if they don't get that far, they were almost certainly thinking something like, "The fact that so many users are interacting with G+ will surely convince some investors that Google could be a significant player in social media, in addition to dominating search."

There's a big reason corporate leaders at large tech firms are so concerned with what investors think of their future prospects. That's because as long as a large tech firm can claim that it is growing, it will enjoy a substantially higher stock valuation than a comparable firm that is "mature."

Let's unpack that: publicly traded corporations all have something called a *price-to-earnings ratio* (or P/E). If a company has a P/E of ten, then for every dollar the company brings in, its corporate valuation is $10 (a company with a P/E of ten that makes $1,000,000 per year will be valued at $10,000,000).

P/Es are not created equal. Consider two companies in similar lines of business, say, a taxi company and an app-based ridehail company. The taxi business has been around for fifty years, is handsomely profitable, and has never had a bad quarter. Meanwhile, the ridehail company is only two years old and it *loses* money every year.

But there's another key difference: their rate of growth. The boring taxi company made a million bucks this year. Last year, it also made a million bucks. As inflation rises, the boring taxi company's annual revenue rises with it, but the earning power of the total revenue of the taxi company doesn't change. Ten years ago, you could buy a million loaves of bread with the taxi company's total revenue. This year, you can also buy a million loaves of bread with the take.

This is called a "mature" company. There are lots of them, and there are good reasons to invest in them—say, if you want a safe bet on a small but steady return. They're the boring, safe stock-market equivalent of a mom-and-pop landlord with a duplex that provides the owners with a home downstairs and a rental flat on top that covers their expenses, year in and year out.

It's a safe bet—unlike the ridehail business, which is operating in a risky field that has seen a lot of failures and just a few successes: Uber and Lyft.

Uber's never been profitable, at least not by the bedrock principles by which investors measure profitability for normal companies. It resorts to the weirdest accounting gimmicks imaginable to make the case that it's actually in the black: for example, many of Uber's foreign operations were such disasters that Uber eventually "sold" them to overseas rivals, who are themselves failing, and who bought out the local Uber operation with illiquid private stock (stock that isn't traded on a stock market and can't be readily converted to cash). Periodically, Uber will declare that their holdings in some dog-shit failing overseas ridehail company have massively increased in value, and, on that basis, declare that it had a profitable year. This is the corporate version of your weird uncle who strains your Thanksgiving dinner conversation by bragging that his holdings of precious Beanie Babies and NFTs have skyrocketed in value and therefore he is now a millionaire.

Lyft, meanwhile, took years to turn even a modest profit, and its actual rate of profit is *very* low, even with the majority of its expenses being borne by drivers who are misclassified as contractors and who supply the vehicles, fuel, and maintenance to keep Lyft's operation running.

So our ridehailing business—only two years old, yet to turn a profit—looks like a worse bet than the taxi company. But there's one way in which the ridehail service and the taxi company differ that makes all the difference in the world: *growth.*

The ridehail service has increased its ridership—and its revenue—tenfold for every year it's been in operation.

Like the taxi company, it is booking a million dollars in revenue per year. Unlike the taxi company—whose profits are $200,000 a year—the ridehail company is *losing* a million dollars a year.

But two years ago, the ridehail company made $1,000. Last year, it made $100,000. This year, it made $1,000,000. It is a *growth* company. And the thing is, it's impossible to predict when that growth will end. Take Uber: in the company's S-1 (the document given to potential investors before a company debuts on the stock market, enumerating its strengths, weaknesses, and plans for the future) from 2019, Uber told investors that it planned to displace *every* ride in a motorized land transport in the *world*: every bus, taxi, streetcar, and subway ride *on the planet*.

If Uber succeeds, then buying a share in the business isn't merely staking a claim on the revenue it's making today—it's a bet that pays out with a share of the incredible, world-swallowing revenue the company will generate later on, as it grows. A share of Google in 1998 was a claim on the nonexistent revenue of a small but well-regarded search engine that lacked any kind of business model. By 2025, it's turned into a share of a company that controls *90 percent* of the world's search traffic, bringing in $350,000,000,000 the previous year.

When investors believe a company is a "growth stock," they are prepared to pay *much* more than they are for a share in a "mature" company. Back in 2020, Tesla hit a P/E of more than *1,000:1*. Around that time, Ford's P/E was less than 10:1.

Why does this matter? After all, valuation isn't the same thing as *revenue*. The company doesn't own any shares in itself—those shares are all disbursed among the original investors, the company's executives, the institutional investors like insurance companies and pension funds, and retail investors like you and me, who might buy a few shares through a Schwab account for our retirement. When the company's share price goes up, the company doesn't get richer, its *shareholders* do.

But here's the thing: when a company has a solid growth stock, other people *want* that stock. The company can use stock to buy stuff, like other companies. Key personnel can be hired at whopping compensation rates, with the majority of that compensation coming in stock.

Where does the company get the stock to buy these companies and hire these genius coders?

By typing zeros into a spreadsheet.

A company's own stock is an *endogenous* substance—it emanates from within the body corporate. There are some limits on how much stock a company can issue without getting into trouble, but the fact remains that *the company issues its own stock*. It doesn't have to rely on any external force to carry off this trick.

Contrast that with a mature company, one that is bidding on the same resources: that mature firm is expected to pay for things with *money*, which is *exogenous* to the company. Money comes from your country's central bank and its fiscal agents (such as chartered banks). Money does *not* come from within a company. If you doubt it, by all means, put this book down, head into the office, and use the company laser printer to run off a stack of U.S. $100 bills (drop me a postcard from federal prison and let me know how the experiment went).

Say our taxi company wants to compete with the ride-hail company. The taxi company's got some cash in the bank, of course, so it can make a job offer to some smart app coders and a project manager to oversee the production of a taxi-summoning app. If they're feeling really ambitious, they can go to their bank manager and ask for a loan so that they can buy a little app startup, or they can even approach an outside investor to buy a share of the company with funds that can be used to bid on those key personnel and acquisitions.

But the ridehail company *also* wants to hire key app programmers and buy promising app startups. Unlike the taxi company, they don't have to dip into their (nonexistent) savings to make these purchases: they can offer stock instead, which they generate by typing zeros into a spreadsheet. Not only that, but they can approach their own bank manager and take out a loan, and *stake their stock as collateral*, securing a preferential interest rate that's better than anything the taxi company can get (the billions Elon Musk uses to buy companies like Twitter and institutions like the U.S. presidency are all loans, collateralized with

Tesla stock, which is why he's so vulnerable to fluctuations in Tesla's share price). They can also offer a stake in their company to a new investor—who will accept far *less* stock for a far *higher* price.

In other words, growth companies find it far easier to *grow*. They can hire key personnel and buy key firms at a price that's *far* lower than the price paid by their mature rivals. Being a growth company is awfully nice.

But it's also awfully precarious. In finance, they speak of Stein's Law, which holds that "If something cannot go on forever, it will stop." Companies that grow at a fierce clip make for attractive investments—but they also demand close attention on the part of investors, because the faster a company grows, the faster it will reach some factor that *ends* its growth.

Which brings us back to Google: the company commands a 90 percent market share in search. That means that virtually *everyone* uses Google for search. The 10 percent who don't use Google are almost certainly people who've made a conscious, firm choice to use something else. Remember: in 2024, Google was convicted of operating an illegal monopoly because of the tens of billions of dollars the company spent every year, buying up default search status on every operating system, device, and browser. It's very hard to even *find* a search box that isn't wired into Google's servers, and even if you do somehow discover a search engine you prefer to Google, you will have to change your defaults in every browser and device you use, and remember to do so every time you get a new phone or a new laptop, or just install a new browser. Ninety percent of the market and true systemic ubiquity is—or should be—the very definition of a "mature business."

But what it really means is that Google's growth is unlikely to come from signing up new searchers. I mean, sure, they could try raising another billion human beings to adulthood while convincing them to be Google users (you may be familiar with this project, which Google calls "Google Classroom"). This might just work, but it's gonna take more than a decade, and markets are fickle and impatient.

Google *can* try to juice their search revenue without adding more search users. The company lost *another* monopolization case in 2024, over one tactic they deployed to extract more revenue per searcher: according to records in *US and Plaintiff States v. Google LLC*, Google executives deliberately made their search results *worse*, in the expectation that users would have to repeatedly search Google to get the information they were seeking. Every time you search anew, that's another chance for Google to show you more ads.

But this is a gimmick. Google might be able to juice its search revenue by making you search two or three times to get to your answer. They may be able to double the number of ads on every page. But they can't make you search *ten* times more to get the information, and they can't make you wade through *ten* times more ads. Eventually, you'll get so fucked off with Google's enshittification that you'll find some other search engine to try.

For Google to maintain a credible story about how it will continue to grow, it has to find new lands to conquer. That's why Google was so desperate to make Google+ happen: if they could convince Wall Street that they were a credible competitor to Facebook, then the market would treat Google as though it might conceivably double or triple in size, and value its stock accordingly.

What's more, that sky-high share price will let Google buy the companies and hire the personnel to make that growth possible, poaching Facebook's best product managers and coders, buying up buzzy new social media start-ups (just as Facebook bought Instagram and WhatsApp) and folding them into Google's social media offerings.

Which is to say that if the market believes you can grow, the resulting P/E ratio provides the resources you need to turn that belief into reality.

But this is a double-edged, razor-sharp sword. The corollary of the idea that "a growing company is worth several times more than a similar, mature company" is "once a company stops growing, it becomes vastly overvalued, because it is now a mature company."

If you're holding a lot of stock in a growth company, you can certainly enjoy the ride up, but you need to sleep with one eye open and one fist poised over the sell button. The minute the market decides that the company's growth has petered out, there will be a mass sell-off, and not just because investors have lost confidence in the company's growth prospects, but because investors believe *other investors* have lost confidence in the company's growth prospects.

It doesn't matter if you think the company might keep growing: if no one else believes that, then the price of the shares you're holding are going to plunge, and as they do, the company will lose one of the key factors that will help it grow—namely, that high P/E ratio.

This is why extremely profitable Big Tech companies experience flash crashes in their share price whenever there's a hint of bad news. A recent, spectacular example came in January 2022, when Meta warned investors that it had experienced *less growth* among U.S. users than projected. Within twenty-four hours, investors had staged a panicked mass sell-off totaling $230 *billion*. At the time, this was the largest one-day devaluation ever experienced by any corporation in human history.

But Meta quickly lost its all-time-loser title. In 2025, there was the release of DeepSeek, a chatbot backed by the Chinese hedge fund High-Flyer. The chatbot, which runs on low-end chips, attained performance benchmarks heretofore associated with AI running on massive server arrays in power- and water-hungry data centers.

Within one day, Nvidia—the leading graphics-card company that makes the specialized processors that the biggest AI companies rely on for their flagship "foundation models"—had lost $600,000,000,000 (two! thirds! of! one! trillion! dollars!) in market capitalization.

These mass sell-offs are panicky, but they're not irrational. Though many companies' share prices recover from these flash crashes, each one is a crapshoot. The higher a company flies, the more likely it is that if it stalls out, it will never recover from its tailspin, because the more a company is growing, the higher its P/E ratio will be and the more its stock will drop once it is perceived as "mature."

As painful as the growth-to-maturity share price crash is for investors, it's even harder on the people who run the company itself. Since the 1970s, stock options and stock grants have played an ever-larger role in executive compensation. These are tax-advantaged: you don't pay tax on them until you sell them, and even then, it's taxed as a capital gain at about half the rate of an executive's wages. And, of course, companies love paying key staff in stock because, as noted, they make the stock right there on the premises with a keyboard and a spreadsheet, while dollars have to come from an investor, a customer, or a lender.

That means that corporate execs at fast-growing companies have a disproportionate amount of their *personal* net worth tied up in their company's shares. From the CEO on down, the management of a growing firm is committed to growth not just because they seek to be prudent overseers of a going concern, but also because they don't want to see their own fortunes cut in half or worse.

This is why tech companies are so fantastically, monumentally fixated on growth. It's not the "growth for the sake of growth" that Edward Abbey called "the ideology of the cancer cell." It's because investors want to make sure that their pension savings and personal riches (and the savings and riches they manage on behalf of others) will grow, and that means that they bet very big on growing companies, and bail out as soon as those companies' growth starts to slow.

This is why we have tech bubbles. It's why Google jammed Google+ into every conceivable corner of every product it offered. It's why Facebook changed its name to "Meta" and spent $60 billion on a doomed metaverse. Sure, Zuckerberg would have preferred that you actually *use* the metaverse, but that was really secondary. The main point was to be sure that investors continued to treat Facebook/Meta's stock as a growth stock, to buy some time while Zuck hunted around for another growth story to tell (which is how I came to be writing a book about AI, and how you came to be reading it).

KPIs are a very powerful tool for selling this narrative. As conservative ideologues never tire of repeating: "incentives matter." An executive who makes the company's employee bonuses contingent on getting you to "interact" with a new technology can turn the entire firm's prodigious brainpower and imagination to the task of producing an impressive and growing number that can be shown to investors in order to convince them that the company is successfully expanding into a new line of business.

The problem, of course, is the curse of Goodhart's Law: when management converts the metric ("how many users are interacting with the hot new technology?") to a target ("you will get paid based on how many users interact with the hot new technology"), then it stops being a useful metric.

That's because employees cheat. Tell a product team that they can double their salary by making a given number go up, and they'll find lots of ways to make that number rise that are far easier and more reliable than getting a bunch of users to try out a product or a feature.

Take Facebook's infamous "pivot to video." In 2015, Mark Zuckerberg and company decided that Facebook needed to grow into YouTube's territory. They wanted to convince video creators to make a ton of video for Facebook, and not just video that overlapped with traditional YouTube fare: Facebook was chasing timely, news-oriented videos that could go viral based on their currency and relevance (and not just because they featured someone putting a lemon up their nose or eating an anvil).

So Facebook juked the stats. They told the media that the popularity of video had exploded on their platform. Publishers who sued them alleged that they tweaked their algorithms so that videos stood a *much* higher chance of being "recommended" to Facebook users, and then they fudged the stats so that a video racked up a new "view" if it stayed on your screen for just a few seconds before you angrily swiped it away.

This "worked." The media industry borrowed billions of dollars and raised billions more in investment capital, mass-fired veteran print journalists, built video production studios, and pumped out an endless scroll of video. Some of it was pretty good, too.

When we tell the story of Facebook's pivot to video, we usually focus on these media companies, which makes sense. After years of Facebook users clicking away from the videos Facebook tried to cram into their eye sockets, eventually Facebook gave it up as a bad job and stopped pushing video. Views and ad revenue for media companies' videos fell off a cliff. Investors pulled the plug on the new "video-first" media firms, and bankers called in their loans. A second wave of video journalists joined their former print colleagues on the breadlines. A wave of bankruptcies swept the media industry.

It's only natural that this very visible outcome takes center stage in accounts of the pivot to video, but they were the *effect* of the pivot to video scam, not the *cause*.

Why would Mark Zuckerberg run this colossal sleight of hand? What was it about dominating video streaming that attracted all this nefarious energy?

In short: a growth story. The audience for the pivot to video wasn't the media execs who fired their reporters and retooled around short-form video. It wasn't the users who tuned in and watched those videos (or didn't, as it happened).

The audience was *investors*. The story wasn't "video is popular on Facebook." The story was, "Facebook—which is losing young users and has saturated its market—has discovered a new field to grow into: becoming the dominant video service. This means that Facebook's revenues can grow by at *least* the amount that YouTube brings in every year (assuming Facebook supplants YouTube), but there's more, since Facebook is targeting a new online video niche, and no one can say just how big that might get."

It might even have worked. If Zuckerberg had been right about the public's appetite for newsy, current-eventsy videos, then the trick he pulled on the media industry might have made all those media companies and their investors very rich, and made Zuck richer still.

But even if it didn't work, it still *worked*. Facebook's P/E ratio stayed high, even as its core business stagnated. Facebook retained the growth company's competitive advantage and was able to snap up dozens of companies and hire more of Silicon Valley's most gifted technologists.

Magicians make a big deal out of never revealing their secrets, though of course, many magicians do, and they make YouTube videos and write books showing exactly how their tricks are done. If you watch the videos and read the books, you'll quickly realize that the reason to keep the method a secret is that most magic tricks have a really simple mechanic, and the "magic" comes from a magician's dexterity, showmanship, and patter.

Magicians have lots of ways to distract and entertain you while they trick you into thinking that you have "freely chosen" a card, and then they lead you around by the nose, convincing you that if the card isn't in their hat, up their sleeves, or in their mouth, it must have disappeared. Learn the method, though, and you'll discover that the magician forced you to choose that card, and the whole point of that long list of places the card *isn't* is to distract you from the place where the card *is*.

The pivot to video was a cheap trick. Zuckerberg chose the victory condition (that is, *more video views prove we are going to grow by hundreds of billions*), defined the metric that proved the victory condition has been met (a "video view" is a two-second, fleeting window that scrolls down the user's screen), designed the system that determined the ability to attain the metric (*the Facebook algorithm will push videos in preference to all other kinds of posts*), and then pulled the world's least convincing rabbit out of his hat.

Now, as George W. Bush said, "Fool me once, shame on—shame on you. Fool me—you can't get fooled again." The simple gimmicks used with Google+ and Facebook's pivot to video eventually wore thin as market analysts wised up to them, so tech upped its game.

Have you ever noticed that when you load a video using any of the major streaming apps (Disney+, Netflix,

HBO Max, etc.), you suddenly have to handle your phone like a photonegative, touching only the bezel at the edge of the device? Just brushing against *any* part of the screen instantly switches you to an unrelated video, while getting back to the video you were just watching takes an agonizing eternity.

This isn't an accident: it's just a KPI, rearing its ugly head.

The major expense for video streaming services is, well, *video*. Streaming services have to either create or license video content, and either way, this costs money. These companies make their money through recurring subscription fees, and viewers don't want to pay every month to watch the same videos over and over again.

To retain users, the streamers have to convince them that the video on the service is worth another month's subscription fees. The easiest way to do this is to commission or license a steady stream of blockbuster programs, the way HBO did in the 2000s, when everyone was talking about the "golden age of prestige television."

This is "easy" in the sense that Hollywood is full of extremely talented people: writers endlessly shopping amazing ideas for great TV shows, as well as actors, directors, costumers, editors, postproduction staff, and everyone else needed to make these TV shows into reality.

But it's *hard* in the sense that all these people expect to get paid, and the very best of them command rather handsome salaries.

For a streaming service hoping to limit costs and retain customers, making or buying an endless stream of new, high-quality TV shows is a losing gambit. Instead, these platforms try to convince their subscribers to watch the shows the service already owns—the vast troves of material that you haven't watched yet because it's old, or obscure, or not to your taste, or because you just missed it the first time around.

To convince you to watch this back catalog, the services build recommendation systems, and—crucially—tout these recommendation systems to investors as the way that they will be able to retain subscribers without breaking the bank on new programming. If investors buy this story—if they treat the streaming services as growth companies—they will keep their precious P/E ratios, and with them, the ability to acquire key companies and personnel, and secure loans at preferential rates. In other words, they'll be able to continue growing.

Obviously, the most salient number in this tale is "How many users stopped paying for a subscription this month?" But the streamers insist that there's another number that's nearly as important: "How often did we successfully recommend a back-catalog program to a subscriber?"

After all, even if subscribers are subscribing *this* month, maybe we'll retain *next* month's subscribers if we can convince them that we have a bottomless well of content the existence of which they never suspected, and which they *love*.

So the word goes down (in the form of a KPI): *Your bonus depends on getting a user to follow a recommendation, and don't think you can fool us by just cycling users to some crappy show—the user has to spend at least ten seconds watching your recommended show for it to count.*

And *that* is why merely grazing any part of your screen while watching a show on a streaming app causes it to switch to some other show, *and* it's why it takes ten seconds to get rid of this new show and find your way back to the show you *wanted* to watch.

You are only the secondary audience for streaming video recommendations. The primary audience is investors.

Which brings me, at long last, to AI.

The tech platforms are desperate to convince Wall Street that you love AI, which is very different from convincing *you* that you love AI. Obviously, it would be nice if you loved AI (just like it would have been nice if you'd watched all those Facebook videos), but for the individuals who stand to make titanic amounts of money in salaries and bonuses from keeping the growth stock story going, that's just a sideshow.

Even if you eventually reject AI so comprehensively that the investment bubble collapses, these individuals who are setting the agenda for their companies will either be long gone (with so much money in hand that they can found a dynastic fortune that stretches down through the ages), or will have found another bubble to inflate.

Avoiding AI is even harder than avoiding video recommendations on a streaming service. Every button you used to click to do something useful has been moved to a different part of your screen and replaced with a button that looks nearly identical, and which summons an AI genie that refuses to be dispelled until you've wrestled with the user interface for endless, agonizing seconds.

I use a stock Android phone, a Google Pixel, their flagship device. In the time since I started writing this book, I've accidentally conjured an AI demon by:

- taking a picture;
- switching apps;
- sending a text;
- replying to a text;
- dictating an email; and
- searching the settings for a way to turn off AI.

Every time I make an AI appear on my screen, a team at Google gets a little KPI score boost, and a process at Google gleefully notes the fact that I have "engaged" with AI. I wouldn't be surprised if these logs were granular enough to note that I have "engaged" with AI six times in a week. At some point near the end of this quarter, a team at Google will comb through these statistics, pick the most impressive-seeming, and make a beautiful series of charts illustrating them: "Users who interact with Google

AI returned to it an average of six times in the first week, and their engagements doubled every week."

Google needs to do this, because the alternative to growth isn't stasis, it's collapse. Without growth, Google will see its share price decline to that of a "mature" company, and then it will lose key employees, miss out on key acquisitions, and its cost of capital will go through the roof. Combine all that with the fact that a failure to grow will devastate the personal finances of every decision-maker at Google, and it's easy to see why they would run things this way.

This is how AI has colonized our economy. As Ed Zitron writes in his "The Hater's Guide To The AI Bubble" from July 2025, seven giant AI firms—Nvidia, Microsoft, Alphabet (Google), Apple, Meta, Tesla, and Amazon—account for 35 percent of the value of *all* U.S. stocks. Nvidia accounts for 19 percent of the value of these seven stocks. Nvidia's entire valuation is based on the fact that the other six companies are spending *hundreds of billions of dollars* on Nvidia's GPU chips, which power their AI data center.

Keeping the growth story alive isn't about one company, or one sector. The entire U.S. economy hangs in the balance.

BUBBLE

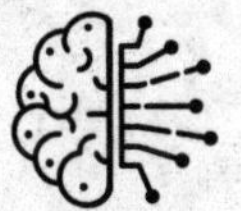

The Funny Math Gets Dangerous

. . . .

Tech companies are spending hundreds of billions of dollars on AI and cramming it into your every orifice with every hour that God sends in a desperate gambit to convince investors that they will conquer the AI market and, in so doing, continue to grow.

For this story to be convincing, investors must first be made to believe that there is some kind of underlying value to AI itself.

Anytime someone pitches a "disruptive" technology as being valuable, what they're saying is that the technology will become valuable at the expense of some existing sector, which it will displace (hence "disruptive"). Amazon's original investor deck told a story of how Amazon could first consume the margins enjoyed by publishers and booksellers (as Jeff Bezos said to the booksellers, "Your margin is my opportunity"). This would give Amazon a chance to launch and shake out its e-commerce service and logistics, so that it could do the same thing to every kind of retailer, and eventually, every kind of manufacturer as well.

Amazon calls this "the flywheel": by offering lots of goods at low prices, Amazon brings in more customers, who attract more sellers, who can be pressured to offer preferential discounts. This in turn creates more goods for sale at low prices, which brings in still more customers and hence more vendors, and so on.

This was a very successful investor pitch, because the investors could be directed to the profits enjoyed by all the companies Amazon planned to disrupt, and do the math for themselves: *If Amazon can take 10 percent of the market, that's x billion dollars; if they get 20 percent, that's 2x*, and so on.

"Disruption" isn't the only way tech companies pitch investors, of course. Sometimes, they pitch them on "innovation," that is, creating a new category. For example, cheap hobbyist microprocessors like the Raspberry Pi and Arduino gave rise to a new kind of electronics tinkerer, who could use them to make everything from home automation to cute homebrew game consoles to advanced electronic art.

But even when entrepreneurs tell a tale about "innovation," they're still usually selling some kind of disruption story: *Uber will make a whole new kind of travel possible—once we bankrupt every taxi operator and kill off all the public transit.*

AI pitchmen make a lot of silly innovation claims

("AI will solve the climate emergency that it has so badly exacerbated by inventing new technologies we can't even imagine"), but the hard-nosed business case for AI is firmly rooted in "disruption."

Specifically, AI sells itself as a way for companies to reduce or eliminate their wage bills by automating tasks that are currently performed by human workers.

When you read through market analyst reports on AI—like the Morgan Stanley claim that AI will be worth $16 *trillion*—what you find is analysts enumerating the kinds of people whose jobs AI can (supposedly) do, multiplying that by the wages all those people earn, and then applying some kind of discount based on what their bosses will pay for software that can do their jobs.

For example: there are about 32,000 radiologists in America, earning an average of $360,000 a year. Morgan Stanley implies that their bosses will be willing to pay 80 percent of that sum for AIs that can do their jobs. Multiply 32,000 by $360,000 by 80 percent, and you get $9,216,000,000, which is judged to be the base market opportunity for an AI radiology app.

That's where the story starts. Next comes an "innovation" multiplier: "Once radiological analysis is too cheap to meter, it will become economical to subject *lots* more people to chest X-rays: *We estimate that hospitals might eventually pay an extra 30 percent for an AI radiologist, once every insured American starts getting an annual preventative chest X-ray.*"

Run these numbers for every job that you think an AI might do, add them all together, and you get those $16 trillion figures.

That gigantic number is the key to the whole AI project. The goal for Google, Facebook, Microsoft, Amazon, and Apple in spending so much to demonstrate that they have a chance of conquering some part of the AI market is to convince Wall Street that they can continue to grow. These companies are *already* valued at trillions of dollars. The only way for them to post meaningful growth is to move into sectors that are *also* worth trillions of dollars.

The tech industry and AI boosters are made for each other. Tech uses all kinds of gimmicks and sleights in proving that users are adopting AI, and AI uses similar gimmicks to prove that AI represents an unfathomably vast market opportunity.

In 2017, just as the current AI bubble was getting started, AI boosters started circulating a story about how AI was coming for the most popular job in America: truck driver. This was an amazing piece of sloppy research compounded by intellectual dishonesty.

It started with the prediction that AI would someday replace truck drivers, something that AI has conspicuously failed to do. Next, these tale spinners consulted the Bureau of Labor Statistics numbers for "truck driver" and determined that "truck driver" was the most popular job in every state. Finally, the AI boosters announced that

America's most popular job was about to be scooped up by AI.

Here's the thing: the BLS does indeed count how many people in each state do a given job, and in 2017, the most popular job in every state according to the BLS was SOC 53-3030, "driver/sales workers and truck drivers." But "driver/sales workers and truck drivers" isn't the same as "truck driver"—that's the category the BLS uses for anyone who drives any kind of delivery vehicle that's bigger than a car.

The vast majority of these people aren't driving eighteen-wheelers with fifty-three-foot cargo trailers; they're driving smaller vehicles in cities: UPS and FedEx vans, moving trucks, water delivery trucks, lumberyard delivery vehicles, and florists' vans.

These vehicles' drivers haven't been replaced by AI, and they *can't* be. Even if the florist's van drives itself to your apartment building, *someone* has to bring the dozen roses to the apartment lobby, press the buzzer, wait to be buzzed in, ride the elevator up to your floor, ring your doorbell, hand you the flowers, and get a signature. A notional driverless florist's van will have exactly the same number of humans on the clock as a van driven by a boring old meat-person.

When AI boosters blithely asserted that truck drivers would "be replaced by AI," they were referring to people behind the wheel of long-haul big rigs. Supposedly, these vehicles are especially amenable to adaptation to full self-driving, provided they are given dedicated lanes on the interstate and are equipped with tools to let each truck wirelessly network to the trucks around it, so they can coordinate their accelerations and decelerations, allowing them to maintain a high rate of speed with minimal, constant following distance.

First of all, congratulations, you've just invented a shitty version of a train.

Second of all, the economic impact of replacing every long-haul truck driver with software is a lot smaller than you might think. Long-haul truckers are already one of the most precarious, low-paid groups of workers in America today. They are routinely misclassified as contractors, which means they aren't entitled to a minimum wage, sick pay, or benefits. Their de facto employers trap them by issuing predatory loans for their trucks, structured so that if a driver misses a single payment, they lose all the equity they've built in their truck, allowing the employer to repossess the cab and resell it to another desperate driver. Wiping out the wage bill for long-haul trucking will have a disgustingly small effect on the cost of shipping goods around the country in fifty-three-foot trailers.

The whole tale of the AI-displaced trucker was designed to sell a story about the gigantic upsides that AI disruption stood to generate. It was the template for a thousand news cycles in the ensuing years, each of which predicted that some vast swath of the workforce was in imminent danger of being wiped out by a chatbot. The fact that the trucker story was flimsy bullshit doesn't set it apart from any of the tales that came later: they're *all* flimsy bullshit.

The fact that these stories are flimsy bullshit doesn't make them benign. The fact that there's a low probability that an AI will be able to do your job doesn't change the fact that *there's a high probability that an AI salesman will convince your boss to fire you and replace you with an AI that can't do your job.*

This is a distinction with a difference. The AI sales pitch is designed to create two opposing teams: on one side, there are the bosses who are replacing their workers with AI and the customers who will consume the products and services these AIs generate; on the other side are the workers who will lose their jobs to an AI.

This is a hard war for workers to win. If someone who is worried about getting lung cancer is told that the only trade-off they'll need to make to get precise, too-cheap-to-meter AI interpretations of their chest X-rays is that yesterday's fallible human radiologist will have to find another way to make their mortgage payments, they'll wish the radiologists luck in their upcoming job search.

But if we frame this accurately, the alliance tips the other way. Even if you stipulate that AIs can catch solid-mass tumors that human radiologists miss, they have their own failure modes, all the false positives and false negatives that get romanticized as "hallucinations" (rather than *errors*, the more prosaic term we use when discussing other technologies).

There's a centaur configuration of radiologist and AI that every patient could have confidence in: we pair radiologists with AIs that offer a second opinion. Where the radiologist and the AI disagree, the radiologist checks the X-ray again to confirm their diagnosis.

This is doubtless an arrangement many radiologists would happily opt for. After all, radiologists are *already* assisted by all kinds of technology, starting with that X-ray machine and the software that displays the pictures it takes, along with all kinds of options for false color, contrast shifting, and other enhancements that help the radiologist keep you alive.

But the centaur radiologist/AI arrangement makes operating a hospital *more* expensive. In this configuration, the hospital pays the same number of radiologists (or maybe even hires *more* radiologists, given that the existing cohort will process *fewer* X-rays now that they're double-checking some of their work), and on top of that, the hospital pays an AI company for a radiology algorithm.

That's the centaur version of the AI radiology story. It's a story about how hospitals can spend more money to improve health outcomes. They do that all the time— like when they add a PET/CT scanner to their radiology department. Companies that make PET/CT scanners make a decent amount of money, but they don't get daily, front-page coverage on every newspaper and lead every newscast for years at a stretch. If Meta or Apple or Microsoft or Google announced that they were poised to enter the PET/CT scanner market, their stock price would barely register a change.

The *growth* version of the AI radiology story is a *reverse*-centaur story. In this version, the AI salesman convinces the hospital to fire nearly *all* of its radiologists and replace them with AI at a fraction of the cost. The surviving human radiologists' job goes from *reviewing X-rays* to *reviewing AI output*. In the jargon of the AI industry, they become a "human in the loop," charged with confirming the judgments the AIs make at a superhuman clip.

This is an essentially impossible task, thanks to well-understood concepts like "automation blindness." When your job is to review something that is usually fine, you eventually lose the ability to spot when it's *not* fine. The TSA proves this every day: they are the world's all-time champions at spotting water bottles in our hand luggage (we forget to remove the liquid from our carry-ons all the time, so TSA agents get lots of opportunities to practice this skill). Meanwhile, when government "red teams" test TSA security by trying to smuggle in bombs and guns, they almost always escape attention. The human sensory apparatus is just not built to maintain vigilance for something that never happens.

The fact that the AI is pretty good at spotting cancer means that a human in the loop that reviews the AI's findings will eventually be lulled into clicking "OK" every time they're shown a diagnosis—and the likelihood of this happening goes up as the volume and cadence of the assessment tasks increase. This is true whether or not the human in the loop really cares about their job and wants to help keep you alive. Automation blindness isn't a moral failing—it's not mere laziness. It's a nearly universal human cognitive limitation.

We already see this happening in "AI-assisted" learned professions. Lawyers keep getting sanctioned for filing AI-generated briefs filled with hallucinated citations to nonexistent cases.

Academic papers are full of AI-generated nonsense, overwhelming the capacity of human reviewers. Springer Verlag, one of the world's largest, most respected scientific publishers, published *Mastering Machine Learning: From Basics to Advanced*, a $169 foundation text for would-be AI experts. The book was filled with AI-generated hallucinations of imaginary technical papers. When contacted for comment by Retraction Watch, a Springer spokesperson called Felicitas Behrendt insisted that the company relied on "full human oversight."

Programmers find spotting chatbot errors especially challenging. Remember, all a chatbot is doing is making a statistical inference about which word it should place after the words it has already typed. That means that the coding errors that chatbots make are as statistically similar to valid code as is technically possible. It's not just that chatbots make coding errors—they make the kinds of coding errors that are maximally difficult for a human being to spot.

Take "slopsquatting." Coders make extensive use of premade code libraries to perform routine tasks. These libraries typically have regular naming conventions. For example, a library for parsing DOCX files might be called "parsing.text.docx," and a library for parsing RTF files might be called "parsing.text.rtf." But sometimes—often—there are inconsistencies in these naming conventions, so the library for parsing HTML is called "parsing.html.text" (and not parsing.text.html).

AIs are reliably tripped up by these errors. After all, a statistical inference engine can only predict a future that is derived from the past—an AI can only predict things that match patterns from its training. So an AI that has been informed of the existence of parsing.text.docx and parsing.text.rtf will confidently output code that calls upon a nonexistent library called parsing.text.html.

That's where the slopsquatter comes in: a slopsquatter can seek out these irregularities in naming conventions and create new, malicious libraries that have the names that a chatbot is likely to hallucinate. These malicious libraries are dutifully compiled into production code, giving the slopsquatter a back door that works on many different products.

The only line of defense here is the programmer, who has to spot these incredibly subtle errors in automatically generated code that can be produced in volumes that dwarf the output of hundreds or thousands of junior coders. If the human in the loop misses one of these errors, it's their fault, not the fault of the chatbot or the boss who bought the chatbot.

This is classic reverse-centaur stuff. The AI's *primary* job is to decrease the wage bill associated with radiography; it is only *secondarily* charged with spotting tumors on X-rays. Installing a radiologist between the AI and the patient allows the hospital to do the former without taking responsibility for failures in the latter. That's the radiologist's job. They are—in the memorable phrasing of the investment bank analyst and writer Dan Davies—an "accountability sink."

Again, science fiction tells us that the most important aspect of a new technology isn't what the machine *does*, it's who it does it *for* and who it does it *to*. The current AI bubble is being driven by large tech firms' need to tell a growth story to investors, which means that this AI *has* to be deployed in a way that reduces wages, because that's the only thing that customers will pay enough for to make all those investors' money back and more besides.

What's more, the mere existence of a widely accepted AI *story* can help suppress the wages of workers, even before AI is deployed. Workers who anticipate their imminent replacement by AI can be bullied into accepting worse conditions, dropping unionization demands, and taking other measures that shift the distribution of a firm's profits from investors to workers.

Bosses *hate* paying workers, more than just about anything else. The history of industrial relations is the history of bosses finding new ways not to pay workers, from stealing their wages (wage theft accounts for one hundred times more in losses than burglaries and shoplifting combined) to misclassifying workers as contractors (which is how Uber and Lyft get away with not paying drivers for the time they spend tooling around looking for a fare) to offshoring jobs to plain old union-busting.

Bosses hate paying low-waged workers, which is why the majority of workers under a noncompete contract in the USA are fast-food workers, whose bosses force them to sign a contract that bans the person working the register at Wendy's from getting a job at a McDonald's drive-through where they'll earn twenty-five cents more per hour.

Bosses hate paying high-waged workers, which is why the biggest companies in Silicon Valley and Hollywood illegally colluded on a "no-poach" pact where they all promised not to offer raises to one another's engineers.

Bosses have fought every attempt to shift their company's profits from capital to labor, tooth and nail, and they are very bad losers. No matter how tiny a win workers wring from their bosses, their bosses will continue to work tirelessly to claw back that gain, whether that's converting employer-guaranteed defined benefit pensions to 401(k)s, or Amazon forcing warehouse workers to queue up for an (unpaid) hour at the start and end of every shift so their pockets can be searched.

This is such a ubiquitous, invisible, universally true fact of our world that it's hard to even notice it—it's like gravity, an invisible, relentless force tugging down on wages. It's so successful that we hear about it only when circumstances briefly give a small advantage to labor, like during the acute phase of the COVID pandemic, when fear of death and mass illness created a temporary labor shortage that drove up wages. Bosses *lost their minds* about this, insisting that "no one wants to work" and demanding a halt to all forms of cash assistance in hopes of tipping the scales back.

All this makes bosses very easy marks for AI pitchmen. Bosses *want to believe*. They yearn for a world in which workers have less power and command a smaller slice of the profits generated by their labor. They dream of workers who are frightened of starvation and do as they're told.

During the Hollywood writers' strike of 2023, I spent a lot of time on the picket line with my fellow writers. I'm not a member of the WGA, but I'm in an affiliated union, IATSE 839, the Animation Guild (I'm part of their writers' group). The vibes on that picket line were *impeccable*.

One afternoon, as I walked the line with a group of writers, one of them said, "You know, you prompt a chatbot *exactly* the way a studio exec gives notes to a writers' room: 'Make me *E.T.*, except it's about a dog, and there's a love interest, and give it a car chase in the third act.'"

I was thunderstruck: this is exactly right. The difference is that writers are bolshie and mouthy, and they're not shy about telling execs when their notes are stupid. Chatbots, on the other hand, are completely pliable: give that prompt to ChatGPT and it will dutifully shit out a whole screenplay in a matter of moments. The fact that it's a *terrible* screenplay isn't all that important: the AI companies assure us that the quality of the writing will only improve, after all. The vital thing about replacing a writers' room with a chatbot is its pliability. For studio execs, getting a quality screenplay takes a back seat to swapping out your disrespectful writers for a lickspittle chatbot that a) works for free and b) does exactly what it's told.

Scratch any AI-driven unemployment story and you'll find a boss whose biggest workaday headache is a worker with human needs. Jeff Bezos refuses to give Amazon warehouse workers and drivers bathroom breaks, which is why there's so much piss in old water bottles on the roads approaching Amazon's depots. Amazon has banned pissing in bottles, but they *also* set schedules that don't have room for a bathroom break, and "just don't piss" isn't an option for workers, so the pee bottles continue to rattle around in the gutters en route to every warehouse.

Whenever anyone asks Bezos a pointed question about the conditions in his workplaces, he changes the subject to the imminent moment in which all of those workers will be replaced by robots. Clearly, he relishes zeroing out that wage bill, but he's also palpably excited about replacing humans (who don't want to starve, or be maimed, or pop a kidney) with machines that are entitled to no moral consideration whatsoever.

As the AI consultant Elijah Clark told Gizmodo's Luc Olinga: "AI doesn't go on strike. It doesn't ask for a pay raise."

Practically everyone who falls for the AI hype is dreaming of getting a human need fulfilled without having to extend moral consideration. That's as true of the AI girlfriend weirdos as it is of the bosses hoping to use AI to replace half their workers and terrorize the remainder. An AI girlfriend is worse than a real girlfriend in the sense that they aren't a person, but that's also what makes them better than a real girlfriend, because unlike a real girlfriend, an AI girlfriend demands nothing of you, except those things you tell it to demand of you.

The difference is that AI "workers" are fueling the AI investment bubble, while AI girlfriends are merely sopping up some of the marginal excesses at the periphery of the bubble.

This is likewise true of virtually every AI service you interact with: AI doodling tools, including the ones that go briefly viral by aping the style of Studio Ghibli animation, AI pornography tools, and AI bots that do your homework.

None of these tools appear on the income projections of the big AI companies. When they commit to spending $100 billion on new data centers and in anticipation of recouping all that capital expenditure and then some, those spreadsheets don't have a line item for "AI doodles drawn at the request of social media shitposters."

Which is not to say that all those peripheral uses have no point. The AI industry likes those uses because they keep people buzzing about AI, which helps their sales agents when they try to pitch your boss on firing you and replacing you with an AI that can't do your job. These peripheral uses also soak up excess capacity in the system and produce a little bit of cash flow—they're the equivalent of the discounted day-old donuts the store puts on sale after five P.M.

We can generalize an important point from this insight: *If you want to puncture the AI bubble, you should train your fire on the applications that are used to justify the massive investment in data centers and training.*

That's not to say that these other uses are benign or even neutral. Getting sexually harassed at school or work by someone making super-realistic deepfake porn of you is ghastly and traumatizing. Scams that use a deepfake to impersonate your boss or your kid or someone else you love can strip you of your life savings and mire you in identity-theft hell for years to come. Political disinformation campaigns that trick unsuspecting people into believing lies are corrosive to our democracy.

These uses all constitute genuine threats that deserve to be taken seriously. However, none of these are the basis for the AI bubble. If we were to regulate all of these uses out of existence and hit anyone who violates these regulations with life-destroying fines, it would have no meaningful impact on the investment prospects for the AI bubble. These are *convincers*, not *business models*.

Never forget that you aren't the target for AI hype—*investors* are. To the extent that anyone thinks of you when designing the publicity and marketing campaigns for AI, they are merely hoping that you will be visibly and loudly impressed.

Being bombarded by messages aimed at someone else can be disorienting. Drive around parts of DC and you'll find yourself passing gigantic billboards advertising obscure elements of the military-industrial complex, aimed at a handful of high-level congressional staffers and military personnel, but on display for millions of randos. Likewise, if you drive 101 or 280 past San Francisco International Airport or the San José Mineta International Airport, you'll pass gigantic electronic billboards, pumping out tens of thousands of ANSI lumens that glow even at high noon, seen by thousands of commuters, but there to pitch only a couple dozen VCs and executives at major tech firms.

Being bombarded with messages aimed at someone else can also be *infuriating*. When my novel *Walkaway* came out, my commonwealth publisher sent me on a tour of Australia and New Zealand, and I took my daughter, who was then ten years old. After half a dozen stops in Australia, we flew to Wellington for the Writers Festival, landing a day early with the intent of doing a little sightseeing. Some local friends took us to the countryside to visit a working farm, and on the way back we stopped at a large grocery store to buy some snacks for the hotel room.

"Don't worry," my friend Daniel said, "we can use the parental checkout aisle." This turns out to be an aisle where they have removed all the sweets that are normally displayed at child's eye height in a bid to inspire kids to whine until their parents buy them a treat.

This was an extraordinary moment—first, because it was so pleasant to have a conflict-free trip through a grocery checkout aisle, and second, because it made me realize how *frustrating* it is to be surrounded by a system that is designed to inspire someone else—your kid—to do something to you that you absolutely want to avoid.

This frustration is just as real whether the person being targeted by the persuasion system is your kid or your boss. And—as every grocery store designer knows—sometimes the most efficient way to convince your real target to take action is to act on someone else, someone with little power or agency.

After the unfortunate affair of the robotic truckers,

the next targets of AI-driven unemployment hype were the commercial illustrators. One day, we all woke up surrounded by an endless wave of AI illustrations, generated by social media users who could prompt a free tool by supplying it with the name of an *actual* artist and a brief description of the picture they wanted to see, and a few minutes later, they'd get a JPEG that fit the bill.

We were told that this was the end of the line for the commercial illustration sector. Every art-school grad was now doomed to a life of (even worse) poverty, and, to make things even worse, the software that was going to put them on the breadline was trained on their own work.

Mira Murati, then the chief technical officer of OpenAI, really got things going when, during an onstage conversation about how OpenAI would affect creative labor markets, she said, "Some creative jobs maybe will go away, but maybe they shouldn't have been there in the first place."

Ms. Murati was either being deliberately provocative here, or very, very stupid.

You couldn't design a better system for provoking outrage if you tried. I refuse to believe that anyone involved in this enterprise was surprised to learn that illustrators were unhappy to be told that they were going to be displaced by tech billionaires who took the fruits of their labor to create their illustrator-obsoleting machines.

That last bit really twists the knife. It's like the People's Liberation Army dragging away your family, executing them, and then sending you a bill for the bullets. It's the kind of thing that Hebrew prophets had in mind when they drafted Leviticus and banned mixing milk and meat on the grounds that it was obscene to cook a baby in its mother's milk.

Here's the thing: as an industry, commercial illustrators *already* make peanuts. When it comes to labor exploitation, illustrators are pretty much the long-haul truckers of the visual arts. Add up all the money that all the magazines and websites stand to save by firing their illustrators and replacing them with AI image generators and you get a sum that won't even scratch the kombucha budget for the gourmet cafeteria at just *one* of the big AI companies.

Remember, you aren't the audience for AI companies' hype. The real audience is the finance sector. To the extent that you are targeted by AI messaging, it's in hopes of getting you to evince some kind of behavior that makes investors think there's something to this AI business.

The fact that AI marketing isn't really directed at AI *users* creates some deeply weird and bad outcomes. In 2023, a bunch of rise-and-grind grifter weirdos came to believe that AI could make them rich by producing *science fiction*, a genre whose average word rate hovers around *zero*.

But science fiction *is* one of the few genres in which there are actual short fiction markets that actually pay something, even if "something" in this case is a couple hundred dollars for a story that you might work on for a month.

One of the best of these markets is *Clarkesworld*, founded by the SF editor and ex–software developer Neil Clarke. Clarke pays 14 cents per word, a very good rate by the standards of science fiction, and runs his publication on a shoestring out of a genuine love of literature (Clarke is a very good editor, and has published many of the field's most beloved and award-winning stories).

So these grifters hear that a) AI is very valuable because b) it can do the work that humans get paid money to do, and they get to thinking. Somewhere out there, there must be a place where we can get paid for the words that ChatGPT will spit out for free.

Before you knew it, *Clarkesworld* was *drowning* in AI-generated "stories," not a one of which even approached being publishable. Why did these dorks drown poor Neil Clarke in so much botshit? Because they'd done the math: at 14 cents per word, they would have to sell Neil *a lot* of words before they could buy a Lamborghini.

These people knew nothing about science fiction, so they assumed that Clarke had an endless supply of 14 centses to pay for the words they emailed to him. They also knew nothing about writing, so they assumed that someone would be interested in reading something they couldn't even be bothered *writing*.

These people made Clarke's life pretty miserable for quite some time, but they didn't make any money. They didn't need to. They just needed to inflate the narrative that AI was coming for human labor.

The material benefit that companies derive by releasing a free tool and ginning up a stampede of furious and terrified illustrators isn't making a ton of money from that tool's outputs: it's that the flood of images helps shore up a story about the coming tsunami of AI-driven technological obsolescence, which would help AI salesmen convince *your* boss to fire *you* and replace you with an algorithm.

While it's important to keep this fact in mind, it doesn't excuse the callousness with which AI boosters dismissed illustrators' concerns, nor does it validate the assertion that AI illustration is an adequate substitute for human illustrations. What it does is help us understand *why* this is going on, and it gives us the analytical and rhetorical tools to undermine the *real* AI story, in which illustrators are a convincer in a big con, not a workforce whose collective salaries would make a difference to the AI companies.

It's not just illustrators, either. The workers who are dead center in the crosshairs of AI bosses are *programmers*. Google, Amazon, Microsoft, Apple—over and over, we hear pronouncements from tech bosses about how many of their coders they plan to fire once the AI works, or (even more ominously) how many coders they've *already* fired because AI works so well.

It's easy to see why AI bosses are so anxious to fire tech workers. For one thing, tech workers are expensive—far more expensive than, say, commercial illustrators. They don't merely draw high wages and demand generous stock grants; they also expect all kinds of perks, from gourmet cafeterias to free dry cleaning and on-site day care (and even free egg-freezing so they can work through their fertile years).

The reason tech workers are able to command all these on-the-job goodies is down to an accident of history: when computers were absorbed into every kind of industrial and personal activity, the supply of trained coders was nowhere near high enough to meet the demand for their obscure, hard-to-master skills. On top of that, tech companies during this period were *insanely* productive, making more than a million dollars per year for every skilled technical employee on the payroll, so anything less than a million per year in total compensation was a bargain.

This meant that coders could demand all kinds of concessions from their bosses, because there were always high-paying jobs with gobs of perks going for anyone who knew how to turn out reliable code on deadline, and bosses could afford to meet those demands and still turn gigantic profits.

But it wasn't enough to get nerds on the payroll: tech bosses had to get them to *work*. And since tech work produced so much value for the firm paying for it, tech bosses chased strategies to get workers to pull the longest possible hours.

One important tactic here was to convince tech workers that they were on a *mission*. That's why these companies adopted such high-flown mottoes, like "Organize the world's information and make it universally accessible and useful" (Google) and "Make the world more open and connected" (Facebook). Tech workers, especially those early workers, were often people whose lives had been transformed for the better by their experiences with networked computers, and many wanted to bring those benefits to the whole world. This made them an excellent audience for this sort of persuasion.

The librarian-theorist Fobazi Ettarh calls this "vocational awe": the process by which employers exploit their workers' sense of duty to the people they serve to get them to accept brutal working conditions. For most professions prone to vocational awe—artists, teachers, nurses, childcare and eldercare workers—those terrible working conditions include low pay. Tech workers got paid a ton, and got to work in whimsical tech "campuses," but they were expected to pull down sixty- and eighty-hour weeks.

Tech bosses labored mightily to stoke tech workers' vocational awe. Part of that shuck required deference to tech workers' conscientious objections to enshittificatory gambits that made the products they worked on worse, in order to make more money for the company. Part of it required that tech bosses approach their underlings as peers and colleagues, enduring weekly company-wide "fireside chats" in which impertinent coders would pelt top management with vicious (but often well-deserved) criticism, which would ricochet around the company's freewheeling internal message boards. In 2018, Google workers by the *tens of thousands* walked off the job, kicking off a series of confrontations that forced the company to abandon a censored search engine for the Chinese market, a $10 billion military project, and its practice of requiring that employees who'd been sexually assaulted by their managers settle the matter in confidential binding arbitration. The exec in charge of the military contract resigned.

Google managers claimed to love their (scarce, valuable) coders, but the instant it became feasible to start firing them, they did. In the space of just a few months, Google declared its first dividend, fired twelve thousand workers (including many of its most senior—and thus most mouthy—technical staff), and declared a $70 billion stock buyback, which would have paid those workers' wages for the next *twenty-seven years*.

Ever since, Google's top managers have issued a steady stream of pronouncements about a) how hard its remaining coders are expected to work ("Sixty hours a week is optimal"—Google cofounder and Alphabet president Sergey Brin), and b) how many coders it will shortly be replacing with chatbots.

Sergey Brin, who returned to the company to oversee its "pivot to AI," has stopped attending the town hall meetings. Mark Zuckerberg has also stopped attending these meetings at Meta, declaring them "not a good use of my time." No one is even *pretending* that Google's top management and its coders are peers anymore. However,

Google is a big ship, with over 180,000 employees. Transforming the company from one in which tech workers get to refuse to work on projects they view as unethical or technically incoherent into one in which workers do as they're told, for progressively lower wages, will require a substantial and sustained act of social engineering.

What are we to make of Google's claim that it can replace coders with chatbots? One strong piece of evidence in favor of this proposition is the chorus of coders on forums and social media who trumpet the incredible efficiency breakthroughs they've experienced thanks to the chatbots in their workflow.

Reading these testimonies closely, it's clear that these coders tend to be in charge of their own workload, pursuing personal projects, or working in a capacity with little oversight. That means that they get to adopt AI tools in a manner and to a degree that they alone decide upon. They're centaurs.

For the reverse centaurs in code shops, it's a very different story. In spring 2025, the *New York Times* reporter Noam Scheiber published a deep investigative dive into life as an Amazon programmer in the age of AI. Scheiber's interviews with Amazon's tech staff painted a very different picture to the one delivered by company CEO Andy Jassy, who touted the ability of AI to save the company "the equivalent of 4,500 developer years."

Scheiber's sources, meanwhile, described how their jobs had been transformed from intellectually stimulating technical problem-solving to a stultifying, frantic race to review the code that Amazon's chatbot had written at superhuman speed, whose bugs were maximally, statistically indistinguishable from valid code. They described how they were laboring under new quotas that couldn't be met *unless* they used AI tools to write the code, which they subjected to the minimum amount of scrutiny to get it to pass the QA tests, and then shoved into production as a new blob of chatbot slop landed on their screens.

In other words, Amazon's coders are now being worked like Amazon's warehouse workers and delivery drivers, among the most abused workforce in the wealthy world.

This shouldn't surprise anyone. Tech bosses have always cultivated a bifurcated workforce, and while they treated their engineering workforce to a degree of chummy pampering that bordered on self-parody, they've always abused the rest of their workers to a degree that *also* borders on (very dark) self-parody. That's true of Apple's Chinese contract manufacturing assembly line workers in the factories of "iPhone City" (girded 'round with suicide nets); it's true of Microsoft's and Google's "green badge" contractors and clickworkers; it's true of Amazon and its bladder-tormented drivers and warehouse workers.

I think that many tech workers genuinely believed that their bosses treated them so well—even as their nontechnical coworkers were so cartoonishly abused—because their bosses *just liked rank-and-file engineers*, especially because so many Google managers started out as tech workers. They respected them, saw them as colleagues, not as workers at all, but rather as founders in waiting, temporarily embarrassed entrepreneurs.

Today, as tech bosses seize on the pretext of AI—along with an oversupply of tech workers—to put the screws to their technical staff, tech workers are learning about the contempt that their bosses have always nurtured for their coddled techies.

But all this still raises the question: Does AI really make coders more efficient?

That certainly doesn't seem to be the case for junior coders, whose "vibe coding" efforts might produce quick-and-dirty tools for one-off usage—for example, a script to reformat a single text file, once—but who lack the depth of experience to vet AI-produced production code for subtle errors that accumulate as "technical debt" in the enterprise systems they work on. Worse: without the hard-won experience of creatively solving a succession of programming challenges, these junior coders may *never* develop the requisite depth of expertise needed to do a thorough review of AI-generated code.

As one Google software engineer told Brian Merchant, whose *Blood in the Machine* newsletter is a treasure trove of firsthand accounts of AI in the workplace, "I have been a software engineer at Google for several years. With the recent introduction of generative AI-based coding assistance tools, we are already seeing a decline in open-source code quality (defined as 'code churn'—how often a piece of code is written only to be deleted or fixed within a short time). I am also starting to see a downward trend of (a) new engineers' readiness in doing the work, (b) engineers' willingness to learn new things, and (c) engineers' effort to put in serious thoughts in their work." Merchant selected this remark as one of fifteen to represent the trends he's seeing in "scores of responses" from workers whose bosses are cramming AI into their jobs.

But what about those senior coders who have the chops to recognize good code when they see it, and aren't shy about telling the rest of us how much effort they're saving with their chatbot "copilots"?

Glyph is a respected senior open-source programmer, founder of Twisted, a widely used networking framework. As both an experienced coder and a founder and overseer of infrastructure-grade software projects, Glyph is an acknowledged expert on how coders approach their work.

In a blog post in August 2025, Glyph proposed a provocative hypothesis about his experienced peers who were touting the miraculous achievements of chatbot coding assistants. Glyph proposed that these chatbots were something akin to a slot machine, in that they mostly did not pay off. Prompt an AI to write you some code, and it doesn't quite work—just as a slot machine that you've fed a dollar to will probably not pay off at all, or might give you a dollar back, or slightly more than a dollar.

This is the expected outcome: it's not unusual for code to break the first time you run it, just as it's not unusual for a slot machine to eat your dollar (or give you back seventy-five cents). Being unremarkable, this event doesn't stick out in your memory.

But sometimes, the AI *does* get the code right: you score a lucky, rare one-shot jackpot, in which your prompt evinces a fully working code solution, which passes all your unit tests and runs perfectly in staging and production.

This is obviously *amazing*, and it looms large in your memory, the same way that a gambler remembers the jackpots with perfect fidelity, while all the losing spins blur into an undifferentiated, low-fidelity slurry. Psychologists and behavioral economists call this "salience bias": things that matter more (like winning a jackpot) are easier to bring to mind. Combine this with another cognitive trap, the "availability heuristic" (we overestimate the likelihood of things we can most easily imagine), and it would be very easy for coders working with LLMs to systematically overestimate how useful these tools are.

But Glyph isn't done. He next inquires about the most common kind of LLM-generated code: code that is neither perfect nor useless, but somewhere in between. For programmers who work with LLMs, it's natural to respond to an output like this by "futzing" with it—tweaking the prompt and/or the code to see if you can fix the bug.

This is like putting another dollar in the slot machine.

People whose gambling is nonpathological have a limited supply of dollars they're willing to feed into the slots. After a few unsuccessful spins, they'll move on.

But for problem gamblers, the only thing that ends their slot-machine session is running out of dollars. They'll keep feeding money into the machine, with their wins standing out brightly in their memory, and their losses fading into the background. If the machine happens to deliver a huge payout, a pathological gambler will think of that as the return on the dollar they fed to the machine for *that spin*, not factoring in all the dollars they lost before the lucky pull.

A programmer who isn't mindful can futz and futz and futz with the LLM, assaying a consecutive, barely noticed series of "Oh, I'll just give this another ten minutes" efforts. If these bear fruit and the LLM produces functional code, an inattentive coder could conclude that after a mere ten minutes, the LLM solved a problem that might have taken three hours to solve without the AI. But they're only thinking about the very *last* ten-minute futz they committed to the project, and it's entirely possible that they spent far *longer* getting the AI to write the code than it would have taken them in the first place.

And of course, the AI code-generation companies charge by the prompt, so for them, this dynamic works *exactly* like a casino owner cashing in on a slot machine (well, not *exactly*: the casino slot machines are regulated). To the extent that an AI company can influence the odds of a "jackpot" (say, by routing a prompt to a more computationally intensive, more advanced model), they can dribble these out on an intermittent reward schedule that any casino boss (or Skinnerian rat-tormentor) would instantly recognize.

So, *does* AI help programmers? Maybe. When the programmers are in charge. And even then, those programmers aren't the most reliable narrators of their own experience with these chatterbox slot machines.

For AI companies to make back the hundreds of billions their investors have entrusted them with, they will have to displace a *hell* of a lot of high-waged labor. That's *displace*, not *augment*. AI companies are selling the replacement of workers with chatbots, but chatbots just can't do workers' jobs. They can help some workers, some of the time, but no one wants that—at least, no one in a position to buy hundreds of billions of dollars' worth of AI products wants that.

To sell hundreds of billions of dollars' worth of AI, you need a *killer* demo.

AI "Art"

· · · ·

No one has ever been wowed by a demo of an AI customer service chatbot,* but plenty of people have had their jaws dropped by an illustration, video, or creative writing sample that came out of an AI tool.

Later in this book, I'm going to talk about whether AI art is theft, but before we get to that question, I want to explore whether AI art is *good*.

Let's start by talking about what *art* is, good or bad. I have a definite view on this because I am a working artist, a novelist with dozens of books to his name.

I've been writing fiction since I was six and selling it since I was seventeen. I also dabble a little in visual art through my collages. In all this lifetime of artistic production, I've come to a conclusion about what art is:

Art is what happens when an artist has a big, numinous, irreducibly complex feeling in their mind, which they

* Apart from some very credulous Air Canada executives.

infuse into some artistic medium—a book, a song, a dance, a painting, a photograph, and such—in the hopes of making a facsimile of that big, numinous, irreducibly complex feeling materialize in the minds of people who experience their art.

In other words, art is a *communicative* act. It is an attempt to *transmit* something that can't be rationally or precisely conveyed, by an objectively very weird (but also very satisfying) means.

When I write a novel, I make literally *millions* of tiny artistic decisions. Most of them exist below the threshold of conscious decision-making, but that doesn't mean they don't exist. Word choices, sentence structures, scenic structures, character traits, dialogue, all of it appears on the page as the result of a vast web of decisions about how to convey that big, numinous, irreducibly complex feeling to you.

Some of these choices are more consequential than others. Some of these novels are more successful—artistically and commercially—than others. But every novel involves a *lot* of these choices, and while you could change many of these choices without fundamentally changing what the novel was *about*, there's a minimum threshold of communicative acts, and if the novel drops below that threshold, it ceases to have anything to say. It ceases to even attempt to embody and transmit that big, numinous, irreducibly complex feeling.

This is true of every medium. As I mentioned, I'm an ardent collagist. I publish a daily(ish) newsletter at Pluralistic.net, and its contents are "open licensed" under a

Creative Commons Attribution 4.0 license, which means that anyone is allowed to copy my articles, distribute them, make new works out of them, and sell their copies or their new works (provided they link back to the original and the license, and note whether they've modified the work).

These newsletter articles are all accompanied by an illustration of some kind, and I want that illustration to be as freely reproducible as the article it sits atop. To generate these works, I compose collages from public domain sources, as well as from visual works that are licensed under compatible Creative Commons licenses.

Collaging turns out to be a very interesting and fun way to illustrate the highly abstract and technical ideas I write about. Even more so, by plumbing historical archives for public domain works (broadly—but not exclusively—works produced prior to 1930), I am able to connect the events of our current place and time with the events of a century ago.

Which is how I came to discover Peter Keppler, the founder of *Puck*, a satirical weekly illustrated news magazine that ran from 1876 to 1918. Keppler wasn't just a magazine publisher, he was a talented and prolific illustrator who drew the majority of the best covers in *Puck*'s forty-two-year run, which overlapped with the major anti-monopoly fights of the Roosevelt administration, when Standard Oil and other corporate octopuses were torn limb from limb by "trustbusters" (if you've ever seen an illustration of a fanged, sinister octopus labeled "Standard Oil," wrapping its tentacles around an Earth studded with oil derricks and such, that's Keppler's).

Keppler's work might as well have been purpose-built for collages to accompany articles about concentrated, technology-driven wealth and power and the callow, sadistic monsters who command it. Indeed, they *were* made for that purpose. And lucky for me, the Library of Congress has scanned an extraordinary number of Keppler's illustrations (mostly oil pastels, but also paintings) at extremely high resolutions.

I have spent many happy hours lovingly tracing the elements of Keppler's drawings, blown up so large that I can see the fibers in the newsprint and the individual strokes of his crayon or brush. Somewhere along the way, something very weird happened: I started to understand Keppler's communicative intent at a level that I had never guessed I would. Keppler's lines, his erasures, his overdrawing and cross-hatchings, they all *say* something, something that can't be put into words, only into brushstrokes. Somehow, by decomposing Keppler's works to their most foundational units, I found myself gaining an insight and appreciation into their totality that I never imagined I'd find.

I'm convinced that this is down to the communicative intent of these individual markings, the fact that each one of them reflects a *choice* by an eminently talented and skilled professional who had devoted many, many hours to his craft, which let him make better art, because it gave him a vocabulary of hand and brush and eye and line that conveys textured subtleties that we, as his audience, may never consciously notice, but that we experience when we see his work.

I recognize in those brushstrokes something akin to the word choices and the million other microdecisions I make when I write; the notes and the silences that my favorite musicians wring out of their throats or their instruments; the framing and composition of the photos that take my breath away.

How many decisions should a work of art embody? There's no virtue to adding *more* labor to a work than is needed to convey that big, numinous, irreducibly complex feelings that art exists to convey—no more than a program should have more lines of code than it needs to run (cue a young Bill Gates, sneering about IBM's quest to build "the world's heaviest airplane").

Some artists have found ways to convey things that are so pithy and sharp that they can take your breath away. There aren't a whole lot of decisions in Duchamp's *Fountain*—a urinal on a wall with the words "R. Mutt" painted on it—but it remains an enduring and significant work. Duchamp's whole sculpture could be boiled down to a sentence so short it could be an AI prompt: "Draw a polished urinal mounted to a gallery wall with 'R. Mutt' handwritten on the forward lip."

But embodied in that short description is a world of profound insight into art, grounded in context and aesthetics, such that millions of people have been moved by Duchamp's piece. It's possible that you could add something to *Fountain* to improve it, but I don't know what it is. It's got everything it needs, and nothing it doesn't.

It's possible that if you fed this prompt to an AI image generator, you'd get something that was just as artistically powerful as Duchamp's original. It's possible that this would produce good art, in other words.

But on average, AI art isn't very good because it doesn't communicate very much. Worse, AI bosses don't seem to understand this. Mark Zuckerberg calls his generative AI image/video tool "EMU," for "expressive media universe." But all that EMU can express is whatever is in the prompt, diluted a million to one with a bunch of pixels that represent a statistical average, shorn of any expressive value the underlying images used to calculate them once embodied.

He calls it "generative," but AI art can't generate the kind of turnover that aerates the aesthetic soil.

Let me provide an analogy. A colleague of mine is a law professor. Prior to the proliferation of chatbots, law students were loath to ask their professors for letters of reference, because these were understood to be a choresome labor. You'd ask your prof to go to all this trouble only if you were damned sure that they thought *very* highly of you.

But as chatbots infiltrated higher ed, everyone—profs and students—understood that the way a letter of reference was produced in today's world was for a prof to feed three bullet points to a chatbot along with a prompt to expand them into a law school reference letter. The chatbot would dutifully inflate these three bullet points with five or six paragraphs of florid nonsense, which the prof could paste onto letterhead and sign.

This has led to an explosion of reference letters, and a concomitant cheapening of their value. Today, when my friend's department has an opening for a postdoc, they are inundated with applicants' reference letters, in such profusion that no one could possibly read them.

You may have already guessed at how they solve this: that's right, they feed these letters *back to a chatbot* and ask it to reduce them to three bullet points.

Obviously, this is very perverse. For one thing, the three bullet points the receiver's chatbot generates are only going to be passingly related to the three bullet points the sender fed to their chatbot. This is a very lossy game of broken telephone.

But even if this could be resolved so that you could be reliably assured that the three bullet points the school received were the three bullet points the prof typed in, this still leaves the question of the paragraphs' worth of florid nonsense.

The chatbot that produces those paragraphs knows *nothing* about the student in question. It *can't*. It is a chatbot—a statistical word-guessing program. It hasn't taught the student, or counseled them, or graded their exams or read their papers. The communicative freight of those extra paragraphs is precisely *zero*. They add no information to the reference letter. They *can't* add information to the reference letter.

Now, obviously, the *form* of a reference letter is itself communicative. The fact that a professor sat down and wrote their own paragraphs of stilted puffery on behalf of a student used to convey something about the prof's estimation of the candidate, because no one would do such tedious work for a bad student.

But automation removes this communicative aspect from the letter as well. Apart from those three bullet points and some minor formatting, the letter is untouched by human hands. The prof may have obliged the student in question because they're the best young lawyer the law school has graduated in a generation *or* they may have keyed in those three bullet points because it only took a second, and the student, while undistinguished, at least didn't stand out in class by being abrasive, stupid, or belligerent.

When an artist prompts an image generator (or a creative writer prompts a chatbot), the *only* artistic communication the resulting piece can yield *is that prompt*. The generative AI doesn't know anything about the big, numinous, irreducibly complex feeling in the artist's head. The AI *can't* know that. It's a next-pixel-guessing statistical machine, not a mind reader. They call it "generative AI" but it can't generate any *meaning*. All it can do is add vaporous filler to the meaning that is contained in a human user's prompts.

And just as the three bullet points are diluted across five paragraphs of nonsense in a reference letter, the AI creator's prompt is diluted across the millions of meaningless decisions that the AI makes about the color of each pixel. None of those virtual brushstrokes can convey anything about the artist's intent beyond that original prompt.

This is why AI art is often dismissed as "soulless," but I think a better term is "eerie." In *The Weird and the Eerie*, Mark Fisher describes eeriness as "when there is something present where there should be nothing, or is [*sic*] there is nothing present when there should be something."

All those virtual brushstrokes in the AI image have the seeming of intention, but nothing is intended by them. They are *eerie*.

But eeriness has a tendency to regress to the mean. The only reason the brushstrokes have the seeming of intent is that we haven't ever encountered brushstrokes *without* intent. Once we become accustomed to brushstrokes without an intender or an intent, the brushstrokes lose their seemingness, and they become empty. That's where the soullessness comes in.

There's a kind of mathematics underlying all of this. Works are artistic to the extent that they a) convey something, that is b) meaningful, in c) the most intentional way possible. An artist in possession of a sufficiently pithy, concentrated meaning can convey that in many, many brushstrokes, or just one, and the resulting piece has an artistry that's something like the power of the meaning divided by the amount of material needed to embody it.

This mathematics is stacked against an AI image generator's output. Unless the creator happened upon a prompt of such powerfully concentrated meaning that it can still shine through after being watered down by millions of irrelevant "decisions" taken by the AI, the resulting piece is going to be full of eeriness that will quickly dwindle to soullessness.

This dwindling takes some time. The fact that we automatically ascribe meaning to brushstrokes on the tacit assumption that every brushstroke was a communicative act creates a cognitive illusion in which we find intention in these unintended pixels, the same way we find faces in clouds. This effect can be quite strong, and I think most of us were genuinely struck by the first wave of high-quality AI illustrations in 2020 (I know I was). But that effect fades, as the realization that there's no *there* there sinks in. AI illustration goes from *striking* to *anodyne* without ever passing through *moving* or *enduring* or any of those other affects that make art so intrinsic to the human condition.

This is why *both* AI art and *corporate* art—like the "Corporate Memphis" illustrations of flat, misproportioned human figures done in solid primary colors—are so often called "soulless." Corporate art isn't supposed to express any kind of *human* feeling; rather, it is supposed to humanize the corporation and create the illusion that this immortal colony organism (a limited liability corporation) is somehow speaking to you. But while corporations are legally "people," and while corporations are at least metaphorically "alive," they have no feelings to communicate. A fry cook at McDonald's might take pleasure in your enjoyment of a Big Mac, but *McDonald's itself* feels *nothing*. It can't feel anything. Corporations have no feelings. The art McDonald's commissions represents a bid by the corporation to dress itself in an artist's skin, to impersonate a communicating human so it can forge an emotional connection with us (contrast this with the very human and communicative decor and art choices made by independent restaurateurs, from diners to fancy restaurants).

That's true of AI art, as well. AI has no feelings, so it has no feelings to transmit. The only feelings that a piece of AI art can convey are the ones that are present in the prompt, which don't amount to much, nearly all of the time.

I'm aware that this can sound like a reactionary argument from the lineage that includes such chestnuts as *a photograph will never be as expressive as a painting* and *sequencing samples of other people's performances will never be as musical as playing an instrument.*

So let me say this: I won't rule out the possibility that, in ten years, I'll look back at this moment and say, "I can't believe I used to think that AI prompts couldn't yield good art" (I also won't rule out the possibility that, in ten years, we'll all look back at this moment and say, "I can't believe we ever thought that AI art could be any good, ever").

For someone to make good art with an AI, they will have to increase the quantum of human communicative intent in the AI's output. That is, they will have to say more to the AI (obviously, they will also have to have something to say).

Even today, every now and again, I'll see something AI-generated online that's striking and even moving, and when I contact the artist to ask about it, they'll say that they produced the work through many, many rounds of prompting, including searching for, selecting, and uploading tons of reference material to the system. They'll ask the system to produce many variations on their theme and choose from among them, infusing still more communicative human judgment into the output. Often, they will have directly manipulated the resulting work with a graphics tool like Photoshop or the GNU Image Manipulation Program (GIMP). To the extent that this trend continues, I wouldn't be surprised if I someday found myself artistically moved by the output of an AI, because most of the work won't have been performed by the AI—rather, it will have come from the artist.

This is the trajectory of all new artistic technologies. Brilliant photographic portraiture isn't merely the product of mastery over a camera or a darkroom—rather, that mastery allows the photographer to anticipate the likely output of the machine based on how they operate it, and their artistic talent allows them to recognize, capitalize on, and refine the unexpected happy accidents arising from the uncertainty intrinsic to operating such a machine.

There's a separate question about whether the cost of making this kind of art outweighs the benefit, given the energy demands of (some) AI tools. I mean, if we could make beautiful art from the feathers of endangered owls, that wouldn't justify hunting them to extinction to help some artists express themselves. I'll come to this later, on page 195, when I talk about bubbles and their productive residues.

Even if you accept my assertion that an AI and a human creator might someday collaborate to make some art that you'll find worthy, it doesn't follow that AI art is an exception to the rule that AI hype is all bullshit aimed at pumping up the stock price of large, monopolistic tech companies.

First of all, as already noted, illustration just *isn't that big an industry*. Producing commercial illustration on demand is a neat trick, but it's not going to pay for AI, not even if every commercial illustrator and draftsman is replaced with an image generator.

But that's not going to happen (nor are AI-generated illustrations going to make big inroads into the fine art world, beyond an initial blush of gallerist-driven hype sales, similar to the early days of NFTs). The reason to commission, say, atmospheric illustrations for a film storyboard is that it harnesses the artistic and creative sensibilities of a visual artist, so that these sensibilities can feed into the set, costume, and lighting design, as well as the camerawork and the actors' performances.

In theory, a director could describe their vision to an AI image generator in sufficient detail, and go through sufficient iterations, to capture their own vision. If you're sufficiently *auteur*-pilled, you might like the sound of this. But for these illustrations to actually sub in for the artistic contribution of the storyboard artist that the AI has replaced, the director will have to infuse as much intentional communication into their prompts as the storyboard artist used to deliver with their pencils and brushes.

That's a lot of work for the director, whose labor is billed at a much higher hourly rate than a storyboard artist. I'm sure there are directors who'd be delighted to make that trade-off, but they're not going to save any money doing so, and remember, lower labor costs are the underlying value proposition of AI, the thing that justifies the hundreds of billions in investment for capital expenditures.

But at least that's a product that could actually do what it claims to do, albeit with a heroic amount of labor that obviates some or all of its purported savings. But so many of the AI image-generation business stories are genuinely absurd. They're stories that are pitched at people who *don't* make art, about how they can get more work out of the people who *do*. It's pretty disgusting.

AI Has a Labor Issue (Not a Copyright Issue)

• • • •

It's only natural for artists to be furious over AI bros' campaign to replace all forms of creative labor with automation. Indeed, as I've stated, I think AI bros are waging war against creative labor *in order to create outrage*, because the louder we all shout about how AI can replace workers, the more money they can raise from the capital markets and the more credulous, worker-hating bosses they can peddle defective AI products to.

When creative workers' livelihoods are under threat, they reach for copyright. This is because creative workers commonly don't believe that they *are* workers. Instead, they see themselves as small businesspeople, engaged in business-to-business transactions with the firms that are, in reality, their employers.

Those employers *love* copyright. They have successfully campaigned for a two-generation-long unbroken expansion of copyright (and of adjacent rights, such as trademark). Copyright today lasts longer, covers more kinds of work, carries stiffer penalties, and has fewer exceptions and limitations than at any time in history.

This period of dramatic copyright expansion has coincided with an equally impressive expansion in the size and profitability of the entertainment and culture industry. The dominant firms in this industry are larger and more profitable than at any time in history, too.

And, at the same time, the *share* of those profits that go to actual artists—to creative workers—is lower than it has ever been, both in real terms (how many artists can support themselves from their work?) and proportionally (what share of an entertainment company's income goes to the creative workers who make its products?).

This seems like a paradox, but only if you have a specific kind of cognitive impairment that afflicts neoclassical economists: the inability to consider *power* when thinking about *policy*. Neoclassical economics is obsessed with reducing all political questions to economic questions, and transforming all economic questions to equations that can "prove" which policy would be objectively best ("numbers don't lie").

This is a political philosophy called *economism*, and it has dominated our policymaking since the Reagan years. In order to convert all policy questions to equations, we have to discard (or, at least, assign a numerical value to) all *qualitative* aspects of policy, like "Is this person afraid of what another person might do to them if they demand a higher wage?" Or even "How can we make this group of people less afraid of what their boss might do to them if they demand a higher wage?"

The quest to turn every policy question into an equation requires us to incinerate every qualitative aspect of policy so that we can do math on whatever dubious quantitative residue the process leaves behind.

Economism insists that it is unseemly for us to talk about *distributional questions* (who gets what share of the income from an industry) because the only thing that matters is the impact of a policy on *productivity* (how much an industry can produce). Don't fight for a bigger slice of the pie—fight to make the pie bigger!

Obviously, there are places where getting a bigger pie is better (for example, if we can increase the amount of housing in our major cities, that would be good). But if you pay no attention to the distribution, you end up with one group (capital) owning the whole goddamned pie (say, we build a zillion luxury condos that are held empty as safe-deposit boxes in the sky, assets on the balance sheet of offshore oligarchs that shelter no one, even as homelessness multiplies around the foundations of this forest of starchitect megaprojects).

For creative workers, the mechanism of copyright is supposed to work like this: the instant you fix a new creative expression in a tangible medium (recording a song, typing some words, taking a photo, or daubing a canvas with a brush), a new copyright springs into existence. This copyright gives you the exclusive right to display, adapt, perform, or duplicate this work, for your entire life, and then your estate continues to enjoy these rights for *seventy years*. This copyright is *alienable*, meaning you can sell it or give it away, and it is subdividable, meaning that you can sell *parts* of a copyright, like "English publication rights" or "film adaptation rights."

Economism, lacking a theory of power, predicts that the more copyright a creator has, the more money they can make. For example, when copyrights were extended (in 1998) from "life of the creator plus fifty years" to "life of the creator plus seventy years," economists predicted that this would increase the wealth of creators because they would have something more valuable to sell to entertainment companies. Those extra twenty years had some marginal value for a publisher or a studio or a label, and that would be reflected in the price they paid to creators.

This did not happen.

When copyright got another twenty years, creators didn't take the boilerplate contracts from entertainment companies, cross out the clause that said the company was acquiring the rights to their work for "the entire term of copyright," write in "for my life plus fifty years," and then demand an extra $100 as a condition of changing the contract back to the new, longer term of copyright.

There are a couple reasons for this. The first one is, obviously, those extra twenty years of copyright are worth approximately *nothing* to *anyone*, at least at the moment of a work's creation. There *are* a few works that are commercially significant seventy years after their creator's life, but it's a vanishingly small proportion of all the works that are created, and anyone who claims that they can predict today whether a work will still be in circulation in a century or more is delusional.

(When I worked in a used bookstore, we had whole shelves devoted to writers who produced number one bestsellers *every year*, over long and storied careers, who were all but totally forgotten just a few decades after their death, to the point where we stopped accepting their books, even in trade.)

The other reason, though, is that most creative workers have no power. At any moment in time, the number of workers who would like to sell art to an entertainment company is *far* outstripped by the number of works those companies are in the market for. For work that's unproven—produced by someone who isn't already a bestseller—there is an inexhaustible supply of creative works competing with your own, made by people who will accept a lower wage than you.

That's been the case with the creative industries forever, of course, but over the past forty years, this situation has been badly exacerbated by concentration on the *buy* side of the media industries.

Successive waves of mergers and acquisitions have reduced the entertainment sector to a series of cartels: five big publishers, four big studios, three big music labels, two big app companies, and a single company that controls nearly all the ebooks and audiobooks.

What's more, some of these are the *same* company—the largest movie studio in the world is also the world's largest record company (Universal), and the *second*-largest movie studio is also the *second*-largest record company (Warner). They are also vertically integrated: the big three music labels (Universal, Warner, and Sony), which control 70 percent of all sound recording copyrights, are also the big three music *publishers*, controlling 65 percent of all compositions.

So not only are artists atomized and bidding against one another in the hopes of selling to an entertainment company, but the entertainment companies are also *incredibly* consolidated, meaning that we have few options to shop our work to—fewer bidders in the ongoing auction for our work.

For a minute there in the early 2000s, it looked like maybe tech companies would provide a competitive counterweight to these media giants. There was a moment when YouTube was offering rates to creators that competed favorably with traditional TV and music companies, when Amazon was offering Kindle royalties that beat the Big Seven (now the Big Five) publishers, when Napster and its progeny were offering a direct path for musicians to reach fans.

That dream died aborning. Tech—a fragmented, bumptious industry at the start of the century—consolidated through its own mergers, until it met and surpassed the degree of inbreeding endemic to media.

Consolidated tech found it easy to do deals with consolidated media, and hostility gave way to a jolly fraternity of tech-media companies that all agree on paying creative workers as little as possible.

No company epitomizes this more than Spotify, the first really successful, post-Napster, all-you-can-eat music platform. Unlike Napster, Spotify struck deals with the Big Three labels for access to their catalogs, making them equity partners and letting them design the payout schedule for music licenses.

That Big Three–structured deal sets the rate per stream at a homeopathic fraction of a penny, for all companies and individuals that supply music to Spotify. However, the Big Three get a minimum monthly guaranteed payout from Spotify (while independents don't), and because that minimum payment is so large, and the per-stream rates are so low, most of the money Spotify pays to the Big Three isn't an "attributable royalty" that they have to give to a specific musician.

Look at it this way: Spotify pays a zillionth of a penny for every stream, but it guarantees Sony, Warner, and Universal tens of millions of dollars in royalties every month. If you add up all the fractional pennies owed to the Big Three for their songs streamed that month, it doesn't come close to that minimum. The difference between the cumulative per-stream royalties and the guaranteed minimum fee is the "unattributable royalty" that the label can spend however it wants.

That's one way the labels benefit at the expense of workers in this arrangement. Here's another: recall that Sony, Warner, and Universal are both *suppliers to* and *owners of* Spotify. That means that they can make money in two ways from Spotify: The first is in the form of license payments, which they get as suppliers to Spotify. The second is in dividends and capital gains they receive as co-owners of Spotify.

The difference is that the money they get as *suppliers* isn't theirs to do with as they will. Universal didn't *create* the music it licenses to Spotify; rather, they have licensed it, too, from musicians. Contractually, those musicians are entitled to a share of every cent their music brings in.

But when the Big Three take money out of Spotify as *owners*, they have no legal obligation to share that booty with the creative workers whose music Spotify gets all that money from streaming. The labels can keep some of that money, or all of it, and which musicians they share it with—and on what terms—is entirely up to them.

So it's obviously better for the labels to get money out of Spotify as owners than it is as suppliers. But here's the kicker: the more money they take out of Spotify as suppliers, the less money they can get out of the business as owners. As suppliers, they want Spotify to pay as much as possible for the music it licenses, but as owners, they want Spotify to pay as little as possible for the key inputs that generate money for the business.

This is an irreconcilable conflict of interest, and for the labels, it cashed out in favor of their shareholders, at the expense of musicians. And not just the 70 percent of musicians whose work Universal, Warner, and Sony control—*all* musicians, including independents who do direct deals with Spotify, because those musicians are forced to take the deal the Big Three structured, without the equity stake, the minimum monthly payments, or any of the other goodies the Big Three get (like free publicity and inclusion in top playlists).*

The point of all this isn't merely to point out that giant entertainment companies abuse the creative workers that supply their products—it's to demonstrate that entertainment and tech companies routinely cook up arrangements among themselves that are designed to suppress the wages of creative workers and divide the difference among themselves.

It's easy to look at the epic struggles between tech and content, from Napster, to Viacom's lawsuit to shut down YouTube, to the vicious battles over the IP chapters in international trade agreements like ACTA and TPP, and conclude that tech and media are on opposite sides of an

* For more on Spotify, see *Chokepoint Capitalism*, the book about creative labor markets that Rebecca Giblin and I published with Beacon Press in 2022.

eternal war—and then to view the AI fight as the latest salvo in that long, long fight.

But tech and media aren't *exactly* enemies—they're more like an axis. They absolutely disagree about how much of a creator's wages each is entitled to, but they are 100 percent in agreement that creators' wages are there for the taking. That is how every single one of these tech-versus-media copyright battles has ended, and it's naive to assume that the AI fight will end any differently.

Think of it this way: a company like Getty Images' major expense is paying photographers for their photos. They've done everything they can to lower this expenditure, mostly by buying up their direct competitors so that photographers have fewer places to sell their work, which lets them extract greater concessions from their workforce, getting more rights for less money.

When Getty sues Stability AI for scraping its archive to train an image generator, their position is *not* that AI shouldn't be used to generate images that displace the photos taken by working photographers. Their position is that Getty (and not photographers) should be compensated for the use of those photos to train the model, and that the resulting model should be designed to Getty's specifications (just as Spotify was designed to the Big Three labels' specifications).

Getty has means, motive, and opportunity to replace photographers with image generators. Like every business, they *hate* paying their workforce and relentlessly hunt for ways to reduce their labor expenditures. If AI-generated images can substitute for Getty's current wares, at a lower cost (because software licenses are cheaper than photographers), then Getty will *absolutely* get into the AI-generated image market.

For Getty, a successful suit against Stability AI will mean a cash windfall for Getty (which will charge Stability AI or some other AI company tens of millions of dollars for training access to its images) and a durable commercial advantage thanks to its ability to contractually demand that any image-generating model that is trained on Getty's images must be distributed with "guardrails" that prevent it from producing images that compete with Getty's AI-generated images.

But none of this will benefit photographers! Sure, Getty may toss around a little money after receiving a cash judgment as a PR goodwill exercise, but structurally, Getty is every bit as interested in changing the world so that the photographs it sells cost it nothing as Stability AI is.

Media companies aren't shy about this, either. When Disney and Universal announced a lawsuit against Midjourney over its image generator, the Recording Industry Association of America (RIAA) spammed its entire press list (which, for reasons I don't entirely understand, includes me) with this statement:

"There is a clear path forward through partnerships that both further AI innovation and foster human artistry. Unfortunately, some bad actors—like Midjourney—see only a zero-sum, winner-take-all game. These short-sighted AI companies are stealing human-created works to generate machine-created, virtually identical products for their own commercial gain. That is not only a violation of black letter Copyright law but also manifestly unfair. This action by Disney and Universal represents a critical stand for human creativity and responsible innovation."

The statement was signed: "RIAA Chairman & CEO Mitch Glazier."

I'm going to address Glazier's claims about "black letter Copyright law" [*sic*] in a moment, but before I do, I want to call your attention to a few salient facts about this letter.

Let's start at the beginning, the statement's opening salvo: "There is a clear path forward through partnerships." The RIAA isn't saying that no one should train an AI on Disney's and Universal's materials—they're saying that Disney and Universal should be *paid* when this happens.

Disney and Universal are two of the companies that were at the center of the creative workers' AI strikes in the summer of 2023, because both companies wanted to alter their union contract to allow them to substitute AI for actors and writers, in order to pay both less. These companies do not hide their desire to swap salaries for software licenses.

And the RIAA's top members *include* Disney and Universal. Universal is the country's second-largest movie studio—and it's *also* the largest record label in the world. Warner, the second-largest record label in the world, is also the country's third-largest studio (and so on). So when the RIAA speaks on behalf of its members, it's representing the film industry as much as it is speaking for the record industry. This should be read as a position statement by the movie studios about AI training, in other words, and that statement is, "We are fine with training AI to make models that can displace our workforce, so long as it's on terms that we dictate."

Before we move on from this press statement, I want to address the *last* line on the statement, the signature: "RIAA Chairman & CEO Mitch Glazier."

Who is Mitch Glazier? He's a former congressional staffer who left the government in 1999 after he was caught slipping a copyright-modifying amendment into an unrelated piece of must-pass legislation. This clause changed copyright law to eliminate "termination of transfer" for music.

What's "termination of transfer?" It's a part of U.S. copyright law that allows creative workers to cancel their copyright contracts after thirty-five years. Even if you sign a contract with a label (or a publisher, or a studio) that says you're signing away your copyright for the full term of life plus seventy years, termination of transfer lets you file some paperwork with the U.S. Copyright Office and get your rights back.

Termination of transfer has been game-changing for

creators who were desperate and/or naive enough to sign very bad contracts early in their careers. Many of these artists—record labels call them "heritage acts," which is a euphemism for "Black musician"—made millions for their labels without ever "earning out" their pitiful advances, and were condemned to lives of grinding poverty despite the commercial and artistic success of their music. Termination of transfer let these musicians claw their copyrights back and either demand reparations and a fair deal from their existing labels, or it let them auction their rights among new labels based on the millions they'd generated over the years.

That's the system that Glazier killed off when he snuck a copyright amendment into otherwise unrelated legislation at the last minute. Congress passed the Satellite Home Viewers Improvements Act without knowing about Glazier's booby trap, but outraged musicians made sure they knew it was there after the vote. They overwhelmed the congressional switchboard with complaints, and these were so righteous that Congress actually convened a special session in order to kill the amendment, and shortly afterward, Glazier left government service, having shown himself to be a sneaky weasel, even by the standard of congressional staffers.

But we know what happened next. Glazier got a new job, as chairman and CEO of the Recording Industry Association of America, where he draws a $1.2 million salary to put his name to press releases condemning AI companies for producing models without paying entertainment companies for their training data.

Now, it's possible that Glazier's view was that the millions that an AI company would transfer to the labels would be equitably shared among their musicians. It's also possible that I'm having an extremely vivid dream in which I'm writing this book, but actually I'm a sixteenth-century bishop napping in the nave of a Gothic cathedral in the lowlands of Europe. I wouldn't give favorable odds on either of these possibilities.

Media companies, like other employers, do not represent their workers' interests, except incidentally. It's true that companies that have *no* money can't pay their workers, but that does *not* imply that when a company gets *more* money it shares it with those workers, or that if a company that finds a way to pay those workers less (or stop paying them altogether) would pass that on and continue to part with a single cent that it didn't have to.

When artists back the media companies' copyright claims against AI companies, they are carrying water for firms that fully support the use of AI to replace creative workers.

They're also backing an extremely weird and implausible (not to mention dangerous) theory of how copyright works.

To the uninitiated (and to Mitch Glazier) it can seem obvious that taking millions of documents in order to train a model is a copyright violation. That's because the naive view of copyright is that it allows someone who produces a creative work to control all uses of that work.

But this isn't true and it has never been true. Copyright law has always afforded a wide sweep of "limitations and exceptions," everywhere in the world. These limitations and exceptions are a feature of every international copyright treaty, too, and have been since 1886, when the first international copyright treaty, the Berne Convention, was passed.

In the United States, some (but not all) of these limitations and exceptions fit into a (very poorly understood) legal doctrine called "fair use." The statutory basis for fair use includes a "four-step test" that judges *can* refer to in order to evaluate a new technology or practice to determine whether it falls into fair use. But the four factors are a non-exhaustive list, and judges can—and do—discard them in order to apply the "rule of reason," which is lawyer-speak for "it would be stupid to rule against this use."

The most famous example of this was the 1984 *Betamax* case, where the Supreme Court legalized the VCR despite the fact that Sony's invention blatantly failed all four steps of the four-step fair-use test. *Betamax* concludes that "any technology that is capable of sustaining a substantial non-infringing use" is legal (this will become relevant in a moment).

How does training a model fit into fair use?

Training a model breaks down into three steps:

- First, scrape (or otherwise acquire) a large corpus of works to use as training data;
- Next, use statistical methods to analyze those works;
- Finally, publish the results of that analysis as software code that applies it to transform prompts into new works.

Is scraping fair use? At this stage, it's pretty clear that scraping—even scraping a massive quantity of works, even if those works are in copyright—is unequivocally not a copyright violation. Every search engine makes a copy of every web page it can find, without taking a license or seeking permission to do so (though some search engines voluntarily ignore documents that are tagged as not scrapable by website operators, in a special file called "robots.txt").

Most of those web pages are copyrighted. Many of them are commercial. Not only that, but in 2014, a court ruled that it was legal to scan *every book ever published*, in order to produce a commercial index of those books.

Scraping is a routine affair. The Internet Archive's Wayback Machine scrapes every web page it can find, and stores and reproduces verbatim copies of multiple editions of these pages, even ones that are explicitly commercial, even when that undermines the commercial interests of the publisher. For example, you can use the Wayback Machine to look up cached copies of articles from *The New*

York Times from the years up until 2011, when *The New York Times* stood up its paywall. If you try to access those pages at nytimes.com, you'll be prompted to pay for access, but the Wayback Machine will give them to you for free.

Scrapers capture copies of corporate websites all the time, to produce records of their old policies and prices, the bios of their executives, and other commercially relevant information that's prone to disappear when a scandal breaks out or when a company alters its terms (it's useful to be able to prove that you bought a "lifetime subscription" to your HP ink when the company decides to cancel that offer and start charging you a monthly fee to continue to use your printer).

In Austria, a scraper named Mario Zechner created longitudinal datasets of prices published by the country's major grocery stores, producing damning and convincing evidence of collusion to fix prices.

So scraping is *good*, actually. The true problem of scraping isn't capturing information without permission, it's overwhelming sites with badly behaved scrapers that trash the servers and their internet connections. That's a valid target for regulation, but even if we end the (real) scourge of badly behaved, voracious scrapers, the *well-behaved* scrapers will still gather information that might end up being training data.

In any event: scraping in order to make a transient copy of a work for the purposes of analysis, indexing, and so on, is legal.

That brings us to the next step in AI training: analysis. Fundamentally, AI training is grounded in *counting things* and then drawing inferences from those tallies. A large language model counts words, counts word pairs, counts phrases, counts how many times a word appears in proximity to another word, counts that distance.

One of the reasons AI output is so unreliable is that this method has no connection to *understanding* the subject the words are describing. We sometimes talk about training an AI as though we were teaching it about the subjects of the documents we've used to train it. But when an AI ingests a Wikipedia entry about types of soup, it's not learning anything about *soup*. It is learning about the word frequency in articles *about* soup.

This can yield a model that produces extremely plausible articles about soup, but that plausibility has nothing to do with its accuracy. An LLM can produce an article about soup that is 100 percent accurate as to the distribution of words and phrases in the article, but the distribution of words in articles about soup is only glancingly connected to the accuracy of an article about soup.

Notwithstanding the fact that this is a poor way to generate *knowledge*, it is also emphatically *not* a copyright violation. Counting the words in an article, or the pixels in an image, or the tones in a song *does not violate copyright*.

This is true even if the things you're counting come from an infringing copy of a work. If you go to a flea market and buy a bootleg CD from a dodgy guy selling them out of a shoebox on a folding table, that CD is illegal. It might be illegal for you to possess. It's definitely illegal for the guy to sell it to you.

But if you take that CD home and listen to it and make a table of the number of adjectives in the lyrics to each of its songs, or the tempo of each, or the length of each, none of those *facts* violate anyone's copyright. You can publish those facts, or use them for your own purpose, without violating copyright law.

Yes, someone might charge you with copyright infringement over the method by which you acquired those facts, but the facts themselves don't violate copyright. They can't. Facts are not copyrightable. Copyright is *strictly* for creative works, not objective facts.

This is why the (extremely sleazy and damning) revelation that Meta downloaded zillions of books from a website that offers infringing files has nothing to do with the legality of the model that Meta produces with the facts it gleans by examining those files. The rights holders behind those books may have a claim against Meta for infringing their copyright by *acquiring* the books, but not for *analyzing* them.

In downloading those infringing books, Meta took a strictly optional shortcut. Meta *could* have just bought all the associated ebooks, or taken them out of the library, or bought the books (new or used) and put them on a book scanner. That's expensive and labor-intensive, but it's a onetime cost, and compared to the overall capital expenditure associated with AI training, it's a rounding error.

In other words, suing Meta for infringing your copyright by downloading your book might get you a onetime payment (and there's nothing I'd relish more than seeing billions of dollars diverted from Meta's shareholders to working writers, because, honestly, *fuck* Meta *and* its shareholders). But it will have no impact on Meta's ability to train a model based on the content of your book.

Indeed, shortly after I finished the first draft of this book, Anthropic—a gigantic AI company with billions in funding—offered a $1.5 billion settlement to authors whose books had been used to train its large language model. Those files were downloaded from a website that collects infringing bootlegs of books, and the liability for downloading those books is what Anthropic was seeking to escape with its cash settlement. Anthropic was *not* stipulating that counting the words in those books violated anyone's copyrights.

Now we come to the *final* step in training a model: taking all the things that were counted during the analytical phase and publishing those findings as a piece of software. This is the model itself.

Copyright-based prohibitions on publishing certain kinds of software exist, but they are very narrow and don't apply here (under Section 1201 of the 1998 Digital Millennium Copyright Act it's illegal to publish code that can bypass an "access control"—this is actually pretty terrible, but it's not in any way related to publishing an AI model).

The reason there are so few prohibitions on publishing software is that software is itself a form of expressive speech, protected by the First Amendment, so broadly, it is illegal—unconstitutional—for a government in the United States to ban certain kinds of software.

The fact that software is a form of speech is *extremely good*. The legal precedent for this arose in 1996, when my friend (and now boss) Cindy Cohn argued a case for the Electronic Frontier Foundation (where I've worked for twenty-four years now), called *Bernstein v. United States*. The *Bernstein* case was about whether the NSA could legally prevent Americans from creating, accessing, and using encryption software (software that scrambles and unscrambles data). The NSA took the position that cryptography was a *munition* (no, really) and prohibited civilian access to encryption that actually worked. Instead, the NSA wanted us all to use a defective encryption system called DES 50 that could be broken in about two hours with about $250,000 worth of equipment.

Under this system, anyone with $250,000 could steal or spoof any data: they could get into your bank account, or get into the bank itself. They could eavesdrop on your private messages, or insert new messages into the stream that undetectably impersonated you. They could forge malicious updates to your computer or phone's operating system (or for the software in your pacemaker or car's antilock brakes).

These were the "crypto wars," when the Clinton administration, police departments, and the U.S. national security apparatus teamed up in an attempt to ban working cryptography to ensure that they could always eavesdrop on their adversaries, foreign and domestic. For the anti-cryptography side, this was a prize beyond all measure, and if that meant that adversaries (foreign and domestic) could eavesdrop on the American public, its institutions, and its businesses, that was a price worth paying.

Obviously, the anti-cryptography forces didn't come out and *say* this. Rather, they maintained that DES 50 was sufficiently secure to protect America from everyone (except cops, feds, and U.S. spooks, obviously).

The pro-cryptography side presented policymakers with strong evidence to the contrary. The EFF cofounder John Gilmore built a custom computer called "Deep Crack" that could brute-force DES 50 in two hours, at a cost of $250,000 (fun fact, Deep Crack currently resides in my home office—aka, my garage—where I use it as a desk. I am typing these words *on top* of Deep Crack!).

Deep Crack was an astounding technical feat and a stinging rebuke to the NSA technologists and lawyers who insisted that DES 50 was sufficient to protect the American people.

No one cared.

Deep Crack barely registered with policymakers or the Clinton administration.

But you know what *did* work? The fact that code is speech.

In 1995, the EFF brought suit against the federal government on behalf of a UC Berkeley computer science grad student named Daniel J. Bernstein. Bernstein had written an encryption program called Snuffle that could scramble data far beyond the capabilities of cops, spooks, criminals, snoops, or other computer scientists to unscramble it. He'd made *working* cryptography.

In the EFF's case, Cindy Cohn argued that the free speech clause of the First Amendment to the U.S. Constitution guaranteed Bernstein's right to publish his working cryptography computer code, on the grounds that computer code is a form of expressive speech. Cohn and her team produced exhibits and experts that showed how computer programmers communicate in code the way that biologists communicate with scientific papers and mathematicians communicate with equations (just as writers communicate in words and painters communicate in paint). The lower court ruled in Bernstein's favor, and the Ninth Circuit upheld the judgment. It was official: code is speech.*

---

* To learn more about the Bernstein case, and Cohn's remarkable career, see her newly published memoir from MIT Press, *Privacy's Defender: My Thirty-Year Fight Against Digital Surveillance.*

An AI model is created in three steps: First, transient copies are made of training data (legal under copyright law, essential to accountability, archiving, search, and more). Then, the elements of the training data are counted (a factual exercise that copyright *can't* restrict, since facts can't be copyrighted—nor would we want them to be).

Finally, the insights gleaned from that counting exercise are published as a literary work: a computer program that mobilizes the inferences made from facts to create new works. That work is protected by the First Amendment, because code is speech.

This isn't an argument about whether the works of AI are *good* or not (as already discussed). Nor is it an argument about whether the use of AI to threaten the livelihood of creative workers is fair.

This is an argument about whether training an AI model violates copyright law, and while it is possible to violate copyright law while training a model, it isn't necessary to do so, and models can be created without violating copyright.

But copyright law isn't some objective, unchanging truth about the universe that was revealed to our legislatures as a guide for living a virtuous life. It's just a *policy*. We made it up. We could change it.

Would changing copyright law protect creative workers? I think the answer is a resounding *no*, and moreover, such a move would necessarily create swathes of collateral damage. It is all downside, no upside.

Ban scraping and you'll put everyone who scrapes for good—to record corporate policies and promises, to record the news before it is censored, to archive the internet—out of business. Google will be the last search engine we ever get, because no one else would ever be able to obtain all the permissions needed to index the web.

Ban the deriving of facts from copyrighted works and you'll lay waste to most forms of artistic criticism. You'll also kill off key accessibility technologies, like tools that catalog and skip over strobe effects in video to protect people with photosensitive epilepsy. On top of all that, you'd be creating a system whereby people can own facts about their work, even when those facts demonstrate their errors and deliberate defects. It's a terrible policy.

Ban the publication of facts gleaned from studying copyrighted works and you'll kill everything from search indexes to computational linguistics studies that trace the rise and fall of words and phrases through history.

And on top of all that, it *won't work*.

Take image generators: one of the things that makes visual artists (rightfully) furious is that the model makers promise their customers that their tools can produce works "in the style of" a painter. Those painters see this (correctly) as a bid to create a tool that competes with them, based on their work, so that the fruits of their hard work, talent, and imagination can be usurped by a soulless corporation that hates them and hates art itself.

So perhaps we could give these painters a new copyright that bans this practice: AI companies aren't allowed to train models on my body of work in order to produce new works in my style.

The problem here is that any famous visual artist has inspired whole *schools* of artists who produce work "in their style." There are far more images "in the style of" Maxfield Parrish or Pablo Picasso than either painter produced in their lives. A model could simply be trained on *these* works and *still* produce works in those styles, without ever having to touch an original work by these painters.

And before you say, "Well, then we could create *another* new copyright, one that allowed painters to prevent other painters from reproducing their style." Sure, you *could* do that, but then you'd have to shut down every art school in the world, because the way you learn to paint is by painting in other people's styles, and the way you create your own style is by adapting, merging, distorting, and rejecting others' styles. Teaching painting without allowing painters to copy other painters' styles is like teaching someone to make music without letting them perform any existing songs.

But maybe you're not famous enough to have inspired a vast field of imitators, but you *are* famous enough that someone is going to try to avoid hiring you while still availing themselves of your work by prompting an AI to make work in your style.

We *could* give you a copyright over your style (notwithstanding that this would be a disaster for creative labor itself). Would that help?

It would not.

Remember, creative workers are selling into markets that are dominated by tiny cartels of unimaginably vast corporations. When we bargain with these companies, they hold all the cards, because five publishers (or four studios, or three labels, etc., etc.) can easily converge on a set of contracting terms to which no exceptions are made, and no matter how objectionable those terms are, any worker seeking access to the audiences these corporations control have to accept the terms the corporations have on offer.

That's the answer to the riddle I posed at the start of this chapter: How is it that copyright has expanded in scope and duration for forty years, and the media industry has grown larger and more profitable over that time, but the share of income going to creative workers has fallen precipitously, both proportionally and in real terms? How is it that artists have ended up with less and less to eat from this vastly expanded pie of entertainment industry revenue created with their labors, even as they were given more rights to those labors?

The answer is that giving creative workers more to bargain *with* is just giving them more to bargain *away*, because the expansion of copyright has coincided with concentration in the media industry. Congress didn't give us more copyrights when we had more bargaining leverage and could have demanded more from our employers as a consequence: the expansion of copyright was driven by our employers cartelizing and uniting behind a demand that copyright be expanded.

Now, they *claimed* that they were expanding copyright in order to improve *our* fortunes, but the instant copyright expanded, our corporate masters all altered their contracts to nonnegotiably acquire whatever new rights Congress had just given "to artists." Lacking any bargaining leverage, we were forced to sign away our new rights to our old bosses, who used these rights to get richer and more powerful and thus able to demand more rights (for us, nominally) from Congress, and then demand that we hand those rights to them. Lather, rinse, repeat.

In a world of giant media cartels, giving creative workers new copyrights to bargain with is like giving your bullied schoolkid extra lunch money. It does not matter how much lunch money you give Junior, the bullies are just gonna take that, too. Give your kid enough lunch money and the bullies will amass a fortune large enough to bribe the principal to look the other way. Keep giving that kid more lunch money and the bullies will be able to afford a global advertising campaign demanding more lunch money for all those hungry kids.

So if we create a new right to control training, that will not give creative workers control over AI training (for one thing, no one creator's exclusion from a training set will have a material effect on a model's performance). It will give *our bosses* control over AI training, because they will amend our standard contracts to demand that we sign away our training rights to them.

This is *already* happening, even though no clear AI training right exists: the contracts used in concentrated industries (like the voice actor contracts for video games) require workers to irrevocably assign exclusive training rights to their creative labor to the company contracting for it.

As Mitch Glazier told us: The media companies don't want to ban AI training. They don't want to ban the use of models to replace creative workers. They just want to get paid for the training and to decide how those workers get replaced, and to ensure that the AIs trained on workers they employed don't deliver a commercial advantage to their rival.

Creative workers won't benefit from this arrangement. If a giant corporation is going to eat you alive, do you care whether that's a media company or a tech company? Does it make a difference if the media company that fires your ass and puts you on the breadline and replaces you with an AI trained on your labor is able to prevent *other* media companies from using that AI?

Here's a rule of thumb for tech policy prescriptions. Anytime you find yourself, as a worker, rooting for the same policy as your boss, you should check and make sure you're on the right side of history. The fact that creative bosses are so obsessed with making copyright cover more kinds of works, restrict more activities, last longer, and generate higher damages should make creative workers look askance at these proposals.

A short coda to this section: if you've read this far, you know that I don't think that scraping the web to train an AI is theft, and indeed, I am broadly in favor of the idea that copyright does and *should* permit technical analysis of creative works without permission from the people who made them.

But if you message Grok, the chatbot that Elon Musk crammed into Twitter, and ask what I think of AI, it will confidently assert that I am an AI critic because I think that training a model on creative works is a form of piracy.

What Should Creative Workers Do About AI?

••••

Let's talk about an answer to AI that your boss will *hate*.

In the summer of 2023, the Writers Guild went out on strike against all the major TV studios. At issue: the use of AI in writers' rooms. Specifically: the use of AI to reduce the number of writers in these rooms, a move that would throw many working writers out of work, and reduce the wages of the survivors of these layoffs, because bosses could threaten them by gesturing at all their former colleagues who were now desperate for a job.

The Writers Guild is a militant union. Just weeks before the AI strikes against the studios, the guild declared unequivocal victory in a *four-year* strike against the big talent agencies. These agencies had merged and remerged until they'd formed a cartel of four giant companies that represented nearly every writer, director, actor, and other creative professional in the film and TV industry.

Having seized this extraordinary degree of market power, the agencies started to screw their clients in order to line their own pockets. They switched their compensation model from a commission-basis (where they got 10 percent of whatever they secured for their clients) to a *packaging*-basis.

In a *packaging* system, a studio puts together (packages) a writer, a director, and a group of actors for a film or TV project, and collects a "packaging fee" from the studio. This is their only compensation; they don't charge any commission to the talent they rep.

That may sound like a good bargain, but it creates a dangerous conflict of interest. Under a commission-based system, every dollar the agent brings in for their client nets a dime for them, so agents try to get as many dollars for their clients as possible.

But under a packaging-fee system, the agent and the studio divide up the production budget among themselves, and the more money the creative talent gets, the less money there is for the agents to take home as a packaging fee.

The predictable result of this shift was that the agent-writer split flipped: writers were making 10 percent of what their agents were taking in on their projects. To make matters worse, the Big Four agencies announced their own studios, with whom they would bargain on their clients' behalf, meaning that your agent was employed by your boss, against whom they nominally bargained on your behalf.

This was totally unacceptable, and the Big Four agencies refused to budge on it. So the Writers Guild held a vote, and on one day—April 22, 2019—all seven thousand members of the guild fired their agents. Over the next four extraordinary years, the writers continued to boycott the major agencies, until each one of them had caved and agreed to halt packaging deals and go back to the commission system.

This was an amazing display of unity in the face of innumerable dirty tricks, like agents offering sweetheart deals to select Guild members to scab against their union. It's all the more impressive as a victory because the Big Four agencies are all owned by private equity funds that extracted hundreds of millions of dollars from them by borrowing heavily, using the agencies themselves as collateral. This is a typical private equity maneuver, and it's why the companies that PE scoops up become so enshittified: they have to hack away at their costs and jack up their prices to service the debt that the new PE owners have saddled them with (these are the much-vaunted "efficiencies" of PE ownership).

Despite the fact that the agencies are all on the hook to make these gigantic payments, the Writers Guild ensured that those dollars wouldn't come out of writers' pockets.

This was the context for the Writers Guild AI strike:

a union that had just won a hard-fought victory over a cartel of gigantic companies that wanted to rip them off.

The writers won every one of their demands. Notably, they did *not* demand a ban on AI in the writers' room: rather, they got the right to use AI as they saw fit, but only under circumstances chosen by the writers themselves, and without any reduction in their wages.

In other words, their bosses tried to turn them into reverse centaurs, and they won the right to be centaurs instead.

I've been in a few Hollywood writers' rooms. I can't imagine using AI in those contexts, but I can certainly imagine lots of ways that AI could be useful, like extracting a list of things that a given character has said in every previous episode, as shot, and providing time code for previous episodes containing specific dialogue. This is a time-consuming process, and it's not a chore anyone enjoys, and being able to get on-demand concordances instantly made to order would be incredibly useful for writers on a series.

The Writers Guild isn't just a union—it's a kind of union that is illegal to create today: a union that conducts "sectoral bargaining." Sectoral bargaining is when a union represents all the workers in a "sector" or industry. These used to be very common, but the Taft-Hartley Act of 1947 banned sectoral bargaining, as well as sympathy strikes (when workers at one company walk out in support of striking workers at another company).

Taft-Hartley was an expensively procured assault on trade unions, and its authors knew their business. By banning sectoral bargaining, they gave a huge advantage to bosses in the class struggle.

Creative workers are especially hamstrung by the ban on sectoral bargaining. Very few of us are actually employees, after all, which means that freelance writers, illustrators, YouTubers, and other creators have to bargain individually every time a publisher sends us a contract (and every time YouTube updates its terms of service).

But creators who think copyright is going to solve their AI problem have *already* committed to changing the law—they want to change copyright law. The problem is, changing copyright law won't protect their interests; it'll just confer an advantage to one set of bosses (media companies) in their fight against another set of bosses (tech companies). Neither one of those opponents is going to share one penny more of the spoils with creative workers than they *absolutely have to*.

The Writers Guild showed us how powerful sectoral bargaining is in fighting AI-based wage erosion. That's why our bosses don't like this solution, and why they want us all to fight for more copyright, which they fully intend to extract from us so they can fire half of us and cut the wages of the rest of us.

Creators who are worried about AI would do well to focus on getting the legal reforms that would let them unionize. But there's a part of copyright law as it exists that creative workers should *defend* in order to safeguard their livelihoods.

Under U.S. law and international treaties, copyrights spring into existence the instant a person creates some kind of tangible, creative work. If you make something that is written down, photographed, or painted, and if that work involves some modicum of creativity, then at that instant of fixation, a new copyright is "attracted" to the work that lasts for your whole life and seventy years more.

In copyright jargon, "Copyright inheres at the moment of fixation of a new work of human creativity."

There's a very important word in that formulation: *human*. Copyright is reserved strictly for works of *human* creativity.

Only works of human authorship are entitled to a copyright, as David J. Slater, the "monkey selfie" guy, discovered in 2018. The fact that Slater put a lot of work into convincing a macaque to point a camera at its own face and press the shutter button doesn't warrant a copyright. Copyright isn't merely reserved for human works, it's reserved for *creative* human works. It doesn't matter how much *effort* you put into making something new. Without *creativity*, you don't get a copyright. You can't copyright a phone book, an alphabetical list of all the colors, or a library catalog (though you can get a "thin" copyright if you exercise significant judgment in selecting facts to compile, for example, "a list of all the colors that don't suck" or "the best books in the library").

On the other hand, even works that require the barest modicum of creativity immediately attract a full-fledged, life-plus-seventy-style copyright.

Spend just a few seconds lining up a photo to capture an angle and subject that strikes your creative fancy and you will get the whole load of copyright over that photo, potentially for more than a century (depending on how long you live after you press the shutter).

However, even if you spend a week setting up the *perfect* camera and lighting configuration to faithfully reproduce a sixteenth-century oil painting (which is in the public domain), you will get *no* copyright over the resulting work. A faithful, verbatim reproduction of a public domain work doesn't get a new copyright.

Likewise, even if you spend months befriending a monkey and teaching it to take a selfie, you'll get no copyright over the resulting image, because the only creativity in the selfie itself (as opposed to the monkey-human interactions) comes from the monkey, and *only humans are entitled to copyrights.*

This has enormous implications for creative workers whose bosses want to replace them with AI. It means that the output of an AI *cannot* be copyrighted. The prompt that a human feeds to the AI is a creative work, and *it* is protected for life and seventy (or a flat ninety years if the human is making a work-for-hire for a corporation). But the output of that prompt is born in the public domain.

Now, if you touch up that work—change a few words, twiddle some pixels in Photoshop—you can create a tiny modicum of copyright over those specific changes. But substantively, the work remains in the public domain.

A public domain work belongs to no one. It can be re-produced by anyone and sold or given away. It can be adapted into new works (which can attract new copyrights: when Jerome Robbins, Leonard Bernstein, and Stephen Sondheim created *West Side Story* from Shakespeare's *Romeo and Juliet*, that act of human creativity attracted a new copyright).

This means that, to the extent that our bosses replace us with AI, they do not own the products they hope to bring to market.

The AI illustrations, the AI articles, the AI songs (and also the AI code)—anyone can reproduce them, for free, without permission. If Getty replaces all its photographers with AIs trained on their images, the resulting catalog of work will be free for anyone to take and sell or give away.

Remember: the reason the AI bubble exists is that companies want to prove that they have growth potential, in order to secure high profit-to-earnings ratios, which will let them buy businesses and hire key employees with stock, rather than cash.

The reason AI is a good candidate for this is that investors are convinced that AI salesmen can convince bosses to fire their workers and replace them with AI, and split the savings between the bosses' shareholders and the AI company that provided the bots.

But the one thing that media companies hate more than paying creative workers is having other people take their output without paying for it. This has been true since John Philip Sousa went to Congress in a fury over musicians recording his sheet music and selling the records, when he said:

> These talking machines are going to ruin the artistic development of music in this country. When I was a boy . . . in front of every house in the summer evenings, you would find young people together singing the songs of the day or old songs. Today you hear these infernal machines going night and day. We will not have a vocal cord left. The vocal cord will be eliminated by a process of evolution, as was the tail of man when he came from the ape.

If media companies are faced with a choice between paying workers and losing the copyright over their works, they will pay creative workers all day long. What's more, even if media companies *do* use AI to assist in making their products, they can only secure a precious ninety-year copyright over those AIs' outputs by having a human alter them. The more human labor they put into an AI-generated work, the more protection they get.

There are very few copyright policies that *only* benefit workers and not our bosses, but *this is one of them*. What's more, the U.S. Copyright Office—an institution that has a nearly unbroken record of siding with media bosses over the creators who generate all those copyrighted works—has repeatedly ruled that works generated by AIs *cannot get a copyright*. They have defended this position in multiple courts, including to the appellate division, and have scored victories every time.

The rule that says AI works can't be copyrighted is a massive boon to artists, and it's one that we should defend with all our might. Unlike a demand for a new copyright over training, this policy does more than give us a new reason to be angry that people are experiencing our art in the wrong way. This is a policy that pays our rent, puts braces on our kids' teeth, and funds our retirement. It is a way around the bullies at the school gate that gets us lunch.

What's more, this is a *pro-centaur* policy. If you're a film editor who wants to use AI to deepfake the eye lines of every extra in a crowd scene, this is fine. No one is going to care whether, for four seconds, the eyeballs of one hundred extras are in the public domain. The fact that anyone can take those pixels has no material impact on the commercial fortunes of the resulting film. You, the creative worker, get another tool in your toolbox, without giving your boss a way to erode your wages.

This is a policy that frees up the use of AI to do the drudge work, while keeping the fun parts of making art for human artists. We paint the painting or sing the song or write the book, and the AI can fix a mis-stroke, remove the tape hiss, or identify all the dialogue associated with a character you're thinking of merging with another one.

That's the future that centaurs should want: one where we, the workers, control the means of production, and where policy does not encourage our bosses to fire our asses and replace us with chatbots.

It's the future that AI boosters *say* they want, when they're scrambling for cover: a future where the machines do the drudge work so we can have more time to be creative. It's not the future they're selling our bosses, where the machines do the art (or write the code) and all the humans are allowed to do is describe and edit the machines' output. AI can be (in the words of Princeton's Arvind Narayanan and Sayash Kapoor), a "normal technology." A thing that creative people and others use, sometimes, when it makes sense, with mixed results, but sometimes good ones.

Actual, Existing AI

••••

One of the decidedly weirder aspects of the AI bubble is the number of participants who profess to be genuinely afraid that they are building an evil, all-powerful godlike creature that might enslave or destroy the human race.

I'm a science fiction writer by trade, and I love a good techno-apocalyptic yarn as much as anyone, but, being a science fiction writer, I am capable of distinguishing serious *predictions* from trenchant futuristic parables that can teach us about the present moment.

These fears (or hopes) of a nascent superintelligence that will spontaneously arise if we just give enough computing power and training data to large language models are absurd on their face. They make sense only if you believe that "being conscious" is a matter of being *really good* at guessing which word is statistically most likely to come after the previous one.

It's true that there is no consensus about what consciousness is, but it's also true that advocates for the idea that consciousness is a matter of guessing words really well haven't offered any evidence for their assertion.

Now, it's interesting that Stanford and its environs have produced so many of these claims about incipient superintelligence emerging from really powerful word-guessing programs. Leland Stanford, founder of the university that bears his name, was renowned as a horse breeder, whose ruthless program of culling and breeding produced some of the fastest horses we'd ever seen (and also turned Leland Stanford into an ardent eugenicist, and made Stanford University a global hub of the eugenics movement).

Stanford hit upon a hugely successful (and terribly destructive and cruel) way to make horses run faster than anyone else. But Stanford never said, "Well, guys, I'm pretty sure that if we keep breeding horses to run faster and faster, eventually one of our mares is going to give birth to a *locomotive*."

A conscious being isn't a word-guessing app that knows more words and has more computing power to guess with. Throwing GPUs and training data at AI isn't going to make a superintelligence (and no horse is going to foal a locomotive).

I'm sure that some of these AI boosters truly believe their own hype. As *Trashfuture* showrunner Riley Quinn puts it, these guys are saying, "Ayyyyyyy eyyyyyyyye," while holding a flashlight under their chin and looking in the mirror and scaring their own pants off.

But the AI bubble is a *material* phenomenon above all else, and the crummy science fiction of AI bros does important work to keep that bubble inflating.

First, there is the corollary of the idea that AI is going to be so powerful that it might destroy the human race: any technology that powerful is also going to be very, very *valuable*. "Sure," the AI pitchman says, "the chatbot can't do your employees' jobs *now*, but once these horses start giving birth to locomotives, there's no telling how many of your workers you'll be able to fire!"

Then there's the way that these outlandish sci-fi stories distract us all from actual existing AI, as we find it in industrial and government settings, where it is, by turns, an expensive disappointment and a significant danger to human life and safety.

Where AI is deployed in commercial settings, it consistently underperforms expectations. During the drafting of this book, an MIT study found that *95 percent* of commercial AI deployments fail, with "no measurable impact on profit." The news sparked a panicked sell-off of AI-related stocks, though whether this is the pin that pricks the bubble remains to be seen. However, the mass sell-off does provide a clue as to the market's "animal sentiments," making it clear that many AI investors believe that they have invested in a stock bubble, and while they're happy to hold those stocks as the market continues to inflate, they are ready to bolt at the first sign of trouble.

The MIT study added systemic context to the scattered anecdotes about commercial AI deployments gone horribly awry.

Take Klarna, an unregulated bank (that is, a "fintech company"). Klarna enjoyed a rush of investment and media attention when it launched its buy-now-pay-later payment system, selling itself as a fee-free way to defer payment for purchases. But as Klarna customers discovered the sky-high fees for missing a payment, Klarna's star faded.

The company revived itself briefly by declaring itself to be "AI-first." The CEO allowed a chatbot to deliver the company's annual results to shareholders, announced his intention to be OpenAI's "favorite guinea pig," and, more substantively, fired the customer service department and replaced them with chatbots.

The next year, Klarna had to go on a "hiring spree" in a bid to rehire its former customer service reps and train new ones, because the chatbots had failed completely to deliver.

Then there's Australia's Commonwealth Bank, which also fired its customer service reps and replaced them with chatbots. A *month* later, it declared the move to be "an error," issued an apology to its former employees, and begged them to come back to work. Press reports describe the chatbot deployment as a catastrophe, with managers dropping everything to answer phones as furious customers, fed up with useless chatbots, demanded to speak to humans.

Meanwhile, a peer-reviewed study in *The Lancet*, a leading British medical journal, described a collapse in expertise by endoscopists who'd been assisted in interpreting colonoscopy results by an AI. These skilled professionals experienced a marked decline in their ability to identify potentially dangerous polyps after just months of working with an AI. This is classic "automation blindness," and bodes very poorly for the idea of using medical professionals as "humans in the loop," overseeing AI-based health diagnostic tools.

On top of all this, there are the long-standing, well-understood problems with algorithmic decision support and decision-making in private-sector hiring, lending, and renting, and public service delivery, policing, and the administration of the criminal justice system.

For more than a decade, critics have assembled a mountain of damning evidence about the inescapable bias in the systems we ask to decide who is eligible for a loan, a job, or parole.

Since the 1950s, computer scientists have used the phrase "garbage in, garbage out" (GIGO) to describe what happens when you ask a computer to analyze flawed data.

A classic example of this phenomenon is the use of "predictive policing" as a corrective for racially biased policing. The theory goes like this: "Our cops are prone to 'unconscious bias' when they choose where to patrol and whom to pull over or stop and frisk. We will feed the objective data about arrests and convictions into a predictive model and it will tell the cops where to go looking for crime, based on the places where crime has been discovered in the past."

Perhaps you've already spotted the problem with this approach (if so, you're a lot smarter than the municipal government rubes who've collectively shelled out billions for this stuff). Let me spell it out for you, just in case: the crime statistics that are used to train the model *come from biased policing.*

After all, cops generally find crime where they look for it. If your local law enforcement has a policy of patrolling Black and brown neighborhoods, and stopping Black and brown drivers and pedestrians for searches, then they will primarily find knives, guns, and drugs in the pockets and glove compartments of Black and brown people, *no matter whether Black and brown people are more likely to have contraband in their cars or on their persons.*

So you *start* with a biased statistical picture of the crimes in your town (garbage in), and then you ask a computer to predict where crimes will occur in the future (garbage out). The computer will tell you to go and find more Black and brown people to search, and some of those people *will* have contraband, which will generate more arrests, which will generate more data used to "refine" the model, which will tell you to do *even more* biased policing next time around.

Lather, rinse, repeat. When you feed a defective AI the defective output of its earlier defective conclusions, it grows *much* more defective, with alarming speed.

In 2016, the Human Rights Data Analysis Group (HRDAG), a nonprofit that uses analytics to track human rights abuses and intervene in tribunals, war crimes trials, truth and reconciliation efforts, and other responses to crimes against humanity, did a study on predictive policing.

HRDAG took advantage of a recent National Survey on Drug Use and Health (NSDUH)—an annual survey considered to be the gold standard for obtaining an objective picture of which Americans use illegal drugs, which drugs they use, and how often they use them. The survey told them where Oakland, California, residents had actually used drugs in 2011.

Then, HRDAG fed the *previous year's* Oakland drug arrest records to a predictive policing model (the one sold by PredPol, a leading vendor) and asked it where it thought the cops should look for drugs in 2011, the year of the NSDUH survey.

By comparing the NSDUH survey's highly accurate picture of Oakland drug usage to PredPol's predictions of drug crime for the same year, HRDAG was able to quantify the bias that was transferred from the original police data to the algorithm that it was trained on.

It wasn't pretty. Not only were the algorithm's predictions for the location of drug crime *more* biased than the *actual* police activity in 2011, but when HRDAG simulated feeding the new arrest data that would result from following these biased predictions back into the model, they saw a *massive* increase in the bias of the following predictions. Even small biases in algorithmic assessments quickly snowballed into gigantic ones when the assessments were put into action, the results of which were used to adjust the model.

This is the coprophagic AI problem: when you feed a bot on botshit, you get something AI researchers call "model collapse," a dramatic reduction in the quality of the guesses the bot makes. It's like the errors in a model's outputs are bioaccumulating through the food chain. Whatever the demerits of AI decision support and decision-making that's based on biased data produced by humans in a state of nature, an AI that's trained on the actions of humans that were fed predictions based on that biased data becomes rapidly, irretrievably, almost *comically* biased.

This means that the existing, widely observed problems with algorithmic bias in lending, hiring, parole, sentencing, predictive policing, and child protection orders will get markedly *worse* as these algorithmic judgments are turned into policy by cops, bosses, loan officers, case managers, parole boards, and judges, all of whom are treated as "humans in the loop."

The problem is that these are the same people who generated the biased data to begin with, and they are charged with approving the output of the algorithms trained on their bad decisions.

These people are just as prone to automation blindness as any other human in the loop. But it's actually worse than that: even if these people notice instances of biased recommendations from a model, they might *approve* of the bias, since it reinforces their own preconceptions.

AI boosters are remarkably cavalier about all this. Here's an illustrative tale of two Googlers: Geoffrey Hinton is a Google AI researcher who was awarded a Nobel Prize in Physics for his work on machine learning. Timnit Gebru is a distinguished computer scientist who was fired from Google after she coauthored a paper that was skeptical of industry claims about LLM capabilities and that raised the real social and environmental problems that AI is producing today.

Gebru is part of a vanguard of AI critics who have sounded the alarm about real-world, present-day AI harms.* When CNN's Jake Tapper asked Hinton about Gebru's concerns about the harms everyday people experience as a result of AI in employment, finance, and the justice system, Hinton pooh-poohed these concerns, saying that they "aren't as existentially serious as the idea of these things getting more intelligent than us and taking over."

---

* Gebru's coauthor on the paper that got her fired from Google is Emily Bender. Bender, in turn, is coauthor (with Alex Hanna) of *The AI Con*, which does an excellent job cataloging these abuses.

At a moment when AI algorithms are dictating who gets hired, who gets a loan, and who goes to jail (as well as who gets bombed in Gaza), Hinton dismisses all of these concerns as not "existentially serious," and tells us that we need to put all our focus on the coming day when one of these prize AI mares gives birth to a locomotive.

This is an awfully convenient dismissal. After all, the majority of AI deployments today are in these manifestly bad and harmful contexts, and these are the reference customers for the ongoing cycle of investment in the AI bubble. That means that focusing on how AI is *really bad* at doing people's jobs could conceivably spark an investor stampede.

At the same time, no AI investor is being shown a prospectus that says, "Our business plan is to create an omnipotent, vengeful deity that will emulsify the human race." An AI company that pledges never to do this isn't leaving any money on the table. Delete the line item that reads "summoning the elder gods and unleashing Armageddon" from the AI company's revenue projections and the bottom line will be untouched.

That makes "runaway AI" a very good subject for AI concerns, since (a) it's not real; and (b) eliminating the possibility of runaway AI has no impact on the revenue projections of AI companies.

Indeed, there are a *lot* of revenue-neutral AI harms that draw a *ton* of focus, from deepfake porn to election disinformation. Unlike "runaway AI," these *are* real harms, but they're not part of the big AI companies' business plans.

No one pitches Morgan Stanley or the Saudi royals on sinking $20 billion into AI data centers by promising giant returns from election disinfo or nonconsensual deepfake porn. To the extent that these *are* businesses, they're very marginal, as criminals and scumbags get heavily subsidized access to AI tools that put their evil deeds in reach.

So by all means, let's put a stop to this wretched nonsense, but let's not pretend that doing so will have any material impact on the ability of AI companies to raise money and engage in all their other harmful conduct, all the while inflating a giant, dangerous investment bubble.

On the other hand, pointing out flaws in the revenue models for the things that AI companies say they *will* make money from is doing the Lord's work, undermining the investment case for the bubble.

Take Google's desire to replace search results with AI. Other AI companies, like Perplexity, say they want to do this, too, but Perplexity doesn't have a 90 percent market share in search, unlike Google, who has lost three consecutive antitrust cases in eighteen months.

Google presently makes nearly all of its money from ads, the bulk of which appear on pages that Google doesn't control, followed closely by ads on Google search results. Google's "crawler"—the scraper that makes copies of every page on the internet it can find—is welcomed in by nearly every web publisher, because appearing in Google's search index means that people can locate and visit your page. Also, because Google has a 90 percent market share, *not* appearing in Google's search index probably means that *no one* can locate and visit your web page.

But as Google shifts from a "search engine" that sends users to other web pages to get the answers they're seeking, to an AI-powered "answers engine" that digests the web and has a chatbot summarize its pages in response to queries, they're breaking this system of mutual benefit.

Under Google's AI plans, websites provide training material that Google turns into AI answers, which appear amid ads that financially benefit Google alone. Publishers are commodity back-end suppliers to Google, bearing all the expense of producing and serving training data and receiving nothing in return.

This is a bad deal for nearly every publisher, even publishers like me. My Pluralistic.net newsletter is "open access," licensed under a Creative Commons Attribution 4.0 license that allows anyone in the world to republish my near-daily essays, for free, provided that they attribute the new publications to me and link to the original and the license. This even applies to commercial publishers— I've had giants like Condé Nast republish long articles of mine as major features without paying me a dime.

I use this licensing model for my editorial work because I'm an activist and I want my work to be read as widely as possible.

But even for publishers with attitudes as generous as

mine, the Google AI model stinks. Remember: I use open licensing because I want to be *read*. Google proposes to ingest my work for free and then *not* show it to anyone. Why would I consent to that?

At time of writing, Google's AI summaries appear to be reducing "outbound clicks" (that is, searchers who go on to visit a web page based on Google's results) by about 25 percent, but Google's avowed purpose is to fully replace clicks to web pages with AI summaries.

Even a 25 percent drop in Google-referred clicks has publishers spitting feathers in rage. It's not hard to imagine a near future in which Google's crawler is blocked from most websites.

Which websites *won't* block Google? Well, there're the ones that Google pays, like Reddit, which gets tens of millions of dollars per year from Google in exchange for permission to scrape its conversations. As a result, Reddit pages play an enormous role in Google's summaries and search results, with the entirely predictable effect that Reddit is now being overrun with AI-generated spam. Reddit has also been rocked by a series of bribery scandals, in which its volunteer moderators are accused of accepting cash payments to allow spam and scams to be featured on high-profile forums that play an outsized role in Google search rankings and summaries.

Which brings me to the other kind of website that welcomes being scraped by AI summarizers: malicious pages, hoping to insert false information into the summaries they produce: conspiracy theories, scams, reputation-laundering lies about the powerful and corrupt. These websites will *absolutely* want to be included in the training data for summarizers, because they don't want to be read by humans (who might sense a whiff of bullshit); they want to be folded into a factual-seeming summary by a trusted company like Google.

Of course, the easiest way to generate this material is to generate it with another LLM, and that brings us back to the coprophagic AI problem. As the web fills up with AI-generated slop, which can be created in volumes that no human author could hope to match, which is designed to be ingested by AI crawlers, which is uniquely welcoming to scraping, the amount of AI-generated content in AI training data will only go up and up and up.

This is a recipe for model collapse.

To summarize, then:

1. Google's AI-generated summaries aren't trust-worthy because of AI's intrinsic, insurmountable "hallucination" problem;
2. Nevertheless, these summaries are apt to take in many searchers, by being reliable enough to induce automation blindness;
3. Meanwhile, replacing search results page links with summaries poisons the relationship between web

publishers and Google, and will result in Google's crawler being blocked from much of the human-generated web;

4. The websites that Google pays for access will become irresistible targets for attacks by scammers and spammers, who will use a combination of bribes and AI-generated mass posting to flood those sites with low-quality information;

5. Websites that welcome Google's crawler in an age of ubiquitous summarization are apt to be AI-generated disinformation and scam sites;

6. Which is a double whammy for Google, whose training data will be both disproportionately composed of deliberately incorrect information and AI-generated, model-collapse-inducing slop.

Any one of these should raise grave concerns about Google's "pivot to AI," but taken as a whole, they make for a damning picture indeed.

While the prospects for AI replacing search are grim, there is one area in which AI is unquestionably useful *and* profit-generating: price-fixing.

The digital age has many opportunities for price-fixing. As the noted communist agitator Adam Smith wrote in his anticapitalist pamphlet *The Wealth of Nations*, "People of the same trade seldom meet together, even for merriment and diversion, but the conversation ends in a conspiracy against the public, or in some contrivance to raise prices."

In Smith's day, price-fixers had to drag their inconvenient meat-bodies to a coffee shop or a Masonic lodge to rig the market. Digitization streamlines this stubbornly physical process by establishing electronic "clearinghouses" that competing firms submit their prices to, and then receive "price recommendations" about how much to charge. Firms that don't follow this advice are subjected to high-pressure cajoling, or kicked out of the club altogether. This kind of price-fixing is at the center of multiple legislative, regulatory, and activist campaigns aimed at businesses that jack up the prices of many of life's essentials, from rental accommodations (RealPage) to meat (Agri Stats) and beyond.

You don't really need AI to do this kind of price-fixing: it's sufficient to repeatedly ratchet prices up by small increments over long timescales in order to increase profits at public expense.

But there are other industries where digitization and AI come together to allow for a much more sophisticated and insidious form of price-fixing, which the industry calls "personalized pricing." Critics have other names for this, of course: when it's done to fix the price of goods, it's called "surveillance pricing." When it's done to fix the price of wages, it's called "algorithmic wage discrimination."

The privacy landscape in America is almost comically terrible. Congress hasn't passed a new federal consumer privacy law since 1988's Video Privacy Protection Act, a law that bans video-store clerks from blabbing about which VHS cassettes you've taken home. This is literally the last technological privacy threat that Congress has bestirred itself to address. The resulting policy vacuum—expensively procured by the commercial surveillance industry—has been filled with every manner of grotesque, easily weaponized spying.

Surveillance pricing isn't the worst consequence of this policy malpractice (that's a stiff competition, but I'd give the prize to how easy it is to dox someone these days, something that's routinely done by bottom-feeding debt collectors, murderous abusive ex-spouses, and online Nazis), but it's the one that you're most likely to get bitten by.

Economists call surveillance pricing "first-order price discrimination," which is when a company uses facts it knows about you, personally, to demand a price that exploits your unique situation. In July 2025, Delta Air Lines announced a partnership with Fetcherr, an Israeli surveillance pricing company, claiming that they would be charging each passenger a different price based on the commercial surveillance data they held on them, which would be converted to a price by means of AI analytics.

Delta—and all the other airlines—already quote a bewildering array of prices to customers, based on all sorts of factors. For example, if you try to book an itinerary without an overnight Saturday stay, Delta—and its rival airlines—will infer that you are on a business trip and charge you significantly more (on the grounds that your employer would rather pay extra to get you home on Friday night or Saturday morning than pay you for an extra day's overtime). Likewise, passengers who book at the last minute are charged a substantial premium.

But this is "second-order discrimination." *Everyone* is charged extra for a flight without a Saturday stay, not just business travelers; *everyone* is charged extra for a last-minute fare, not just people who need to get to a funeral or a job interview or to their kid's college dorm so they can bring her to rehab.

Surveillance pricing makes finer distinctions. Many merchants already practice surveillance pricing, including Amazon, which charges different prices depending on whether you are logged in when you search its website. Note that there's no way to use Amazon's *app* without logging in, making the app a perfect vehicle for price gouging.* Amazon knows that most Prime subscribers start and finish their purchases on Amazon (having already paid for a year's shipping in advance), and many Prime customers have done their own research, comparing the prices Amazon quotes them with the prices quoted to friends who *don't* pay for Prime, concluding that Ama-

---

* Section 1201 of the U.S. Digital Millennium Copyright Act felonizes the act of reverse engineering an app, making this kind of rip-off hard to detect and even harder to prevent. Over the past quarter century, the U.S. Trade Rep used the threat of tariffs to force nearly every U.S. trading partner to adopt similar rules. For more, see my 2025 book *Enshittification: Why Everything Suddenly Got Worse and What to Do About It*, from Farrar, Straus and Giroux (U.S.) and Verso Books (U.K.).

zon is especially hard on Prime customers, sometimes charging them much more than non-Prime customers.

The most pernicious form of surveillance pricing isn't in consumer markets, it's in *labor* markets. The conversion of much of the global workforce into "independent contractors" (through the fiction that hiring someone through an app doesn't make you an employer) opened the door to "algorithmic wage discrimination."

That's when a potential employer uses surveillance data about its workforce to adjust the hourly wage offered to each worker, every time they are offered a job. The groundbreaking work of the legal scholar Veena Dubal studied Uber's use of this practice. Uber has lots of information about its drivers that it can use to adjust the wage it offers them. Most important, Uber knows how selective a driver has been, and how that selectivity has changed over time. Drivers who accept bad rides—rides that are far away and/or don't pay much—are assumed to be more desperate and thus willing to accept lower wages. Meanwhile, drivers who have turned down bad rides are assumed to have more financial options (perhaps they have more savings, or are using several different apps to find jobs) and are offered higher wages for the same work. The goal of the algorithm is to predict the lowest wage a driver will accept.

The inverse of this is Uber's passenger pricing algorithm, which uses AI to predict the highest fare you are willing to pay.

Thanks to copyright restrictions that criminalize modifying apps without permission, there is no legal way for either drivers or riders to build comparison-shopping dashboards that bid out the same worker or job to multiple apps and choose from the cheapest. For riders, that means there's no tool that lets you submit the same ride to Lyft, Uber, and your local taxi company's app and automatically choose the best deal (remember: the prices offered by each app might be for a ride in the same car, driven by the same driver, who is signed into three different apps). For drivers, that means that there's no way to survey the going regional rate offered by rideshare, food delivery, and other gig apps from moment to moment and automatically disregard lowball offers—and it also means there's no app that lets all the drivers in a region automatically decline jobs below a certain wage.

The gig model has slithered out of the rideshare and delivery sector and is now being used to bid on the labor of credentialed professionals. In 2024, Katie J. Wells and Funda Ustek Spilda from the Roosevelt Institute published "Uber for Nursing: How an AI-Powered Gig Model Is Threatening Health Care," a report on hospitals' use of apps to procure the labor of contract nurses, who are hired by the shift to cover staffing shortfalls (hospitals having sliced their unionized staff nurse cohort to the bone, these shortfalls are a normal occurrence).

Contract nurses used to find work through regional nurse staffing agencies. There were several of these competing in each market. However, with the rise of app-based

"Uber for nursing" digital hiring tools, the regional staffing agencies have given way to a cartel of four nurse-hiring apps: CareRev, Clipboard Health, ShiftKey, and ShiftMed.

In Wells and Spilda's report, they describe how these apps buy nurses' credit history from data brokers before offering them a shift, and then reduce the wage on offer for nurses with a lot of credit-card debt. Those nurses, after all, are more economically desperate and thus likely to accept a lower wage.

This is a place where AI can really wring out some intensive wage savings. The goal of this system is to accurately guess the lowest wage a nurse will accept. Because these apps dominate the market for contract nursing nationwide, they get *many* chances to lowball a nurse for a shift, and because the United States has no privacy rules to speak of, the outcomes of these offers can be compared with rich, deep dossiers on nurses' financial lives. Locating the biographical and financial correlates of economic desperation and quantifying their implications for eroding nurses' wages is a perfect task for AI.

The business world loves the idea of surveillance pricing. Economists have long dreamed of a day when every purchase and every labor contract is personalized, precision-formulated to the penny, so that everyone who buys your products pays the maximum they're willing to part with and everyone you pay is given the minimum they're willing to accept.

Economists dress this up in pseudoscientific rhetoric

about efficiency, implying that businesses might offer discounts—or pay premiums—if they know more about the people on the other side of the bargaining table. For example, in Norway, digital shelf tags let grocers lower the price of yesterday's milk.

But there isn't a single AI sales call whose takeaway is, "Buy our AI pricing-bot and we will charge your customers less and pay your workers more." The market isn't betting hundreds of billions on AI because investors are excited about offering *discounts*.

No normal person believes that surveillance pricing is going to lower prices, which is why everyone who buys anything from, or sells anything to, a company that employs surveillance pricing *hates* the whole idea.

This leaves companies in something of a bind. They know their investors love the idea of price-gouging customers and lowballing workers, but everyone else hates this nonsense. That leads to a predictable pattern in which companies announce some kind of "variable pricing," "personalized pricing," or "surge pricing," to applause from stock analysts, which then triggers mass outrage and the company abandoning the plan.

That's what happened in 2024, when Wendy's announced that it was going to "surge price" the drive-through, charging extra for hamburgers during busy times. Just two weeks—and a zillion angry social media posts—later, Wendy's insisted that we'd all misunderstood the plan, and scrapped it.

And Delta? Well, just days after announcing that it would use surveillance pricing for airfares, the company publicly abandoned the plan, saying—as with Wendy's—that we'd all misunderstood and gone off half-cocked.

My feeling is that Delta planned to roll out surveillance pricing, even if they oversold its capabilities. What's more, other airlines are also announcing deals with surveillance pricing AI companies, even as they change the rules for their frequent flier programs to severely punish anyone who buys a ticket from a third-party ticket broker. After all, for the company to charge you more based on your personal characteristics, they need to know who you are, and they can't quote a "personalized" price if you are shopping for tickets on Expedia or Google Flights.

Surveillance pricing is a pretty good example of actual existing AI, as we see it in commercial deployments. It is popular with bosses, unpopular with workers and customers, and exists to redistribute money from the latter to the former. It is investor-pleasing, and pretty efficient and effective.

It's terrible.

That's the thing about actual, existing AI: the lucrative present-day use cases, from drone warfare to wage theft, from empiricism-washing, racist policing to ripping off shoppers, are pretty ghastly.

To be an effective AI critic, you need to strike at the source of AI's power, which is the investment capital it attracts. That investment capital is attracted by speculative stories about how AI will someday do our jobs, and by the reality of AI being used today to rip us off and to automate lucrative human rights abuses at scale.

If you want to stem the tide of AI investment, it's imperative that you don't get distracted by Geoffrey Hinton's hyperventilating about the "existentially serious" risk of "these things getting more intelligent than us and taking over." It's vital that we don't conflate "an AI salesman convincing your boss to fire you and replace you with an AI that *can't* do your job" with "AI will steal your job."

It's essential that we never stop reminding people that the current, actually existing lucrative uses for AI are *terrible* and should be banned.

Absent Indians and Gujarati People Typing

· · · ·

For all that actual, existing AI is mostly doing things that are simultaneously boring and evil, there's also a realm of extremely advanced and amazing AI demos that dangle the promise of an extraordinary future for AI, if only we get out of the way and let tech monopolists consume as much water, energy, and capital as they need.

GM's Cruise subsidiary once filled the streets of San Francisco with "self-driving" cars that turned out to be monitored by 1.5 skilled engineers at all times (and still managed to hit a woman and drag her for twenty grisly feet).

That's not the only example of an "AI" that just turns out to be some people pretending to be an autonomous system. Indeed, the first widely touted intelligent machine turned out to be a hidden person!

In the early 1800s, a flimflam artist named Johann Nepomuk Mälzel toured Europe, wowing aristocrats and punters alike with a machine we today call "The Mechanical Turk." The Mechanical Turk was an "automaton," a clockwork sort of robot, part of a wave of impressive and cunning machines that could perform *amazing* feats. For example, Jacques de Vaucanson's clockwork duck could walk, quack, and *eat*, with the special food it ingested later emerging as realistic turds from its exquisitely sculpted cloaca.

The Mechanical Turk put even the shitting duck in the shade. It was a *thinking* automaton, in the shape of a fiercely mustachioed and turbaned "Turk" sitting over a chessboard. If a human were to set up a chess game on that board and make a move, the Turk would make a counter-move, playing a pretty darned good game of chess, using nothing more than clockwork to reason out its play.

Of course, clockwork can't do this. The Mechanical Turk was a hoax: the Turk was controlled by a little person, pretzeled into the Turk's base, who was able to observe the chessboard using a complex, camouflaged periscope.

While an automaton that could pick up chess pieces and move them via a set of hidden controls is a pretty impressive piece of clockwork, Mälzel reasoned (correctly) that there was a lot more money to be made from exhibiting a thinking machine, rather than a remote control chess-playing machine and an extremely patient little person who played pretty good chess.

When Jeff Bezos created the first major "clickwork" platform on Amazon, he called it "The Mechanical Turk," in a semi-ironic move whose vibe can only be summed up as *ha ha, only serious.* The extremely low-waged workers ("turkers") who work on "MTurk" make pennies for filling in surveys, labeling data, and performing other "human intelligence tasks" (HITs). The customers for their labor are encouraged to view turkers as a *service,* not as *people.* You're supposed to treat MTurk like a crude form of sentient computer: you give it plain language instructions and a credit card, and shortly thereafter, your instructions have been followed and your credit card has been debited. Breaking down complex jobs into HITs means that turkers are *fungible*: you can swap out any turker for any other turker and still get your job done. They're *commodities.*

MTurk—and the original Mechanical Turk—presaged many "impressive" AI launches, products, and demos that followed. For example, Amazon opened a chain of retail grocery stores that you "just walked out of" after choosing your purchases. Amazon claimed that a constellation of "AI cameras" observed you as you shopped and debited your account accordingly.

But in 2024, just before Amazon announced that it was shutting down these stores, reporting by Theo Wayt in *The Information* revealed that the "AI cameras" that observed your shopping and debited your account were just regular CCTVs, wired up to vast call centers in India, where up to three workers observed each shopper, furiously clicking their best guess for what the shopper had put in their basket.

It is so common for "AI" to turn out to be low-paid call-center workers in India, Indian technologists joke that "AI" stands for "Absent Indians."

The fake AI isn't always in India, of course. In 2021, Elon Musk demoed a humanoid robot he called Tesla Bot. After talking about how humanoid robots were destined to zero out the wage bills of factory owners (like himself), Musk invited the Tesla Bot onstage to perform an astonishingly fluid series of dance moves, of the sort no robot had ever been seen to perform before.

There's a good reason that Tesla Bot was able to dance like no robot ever seen. Tesla Bot wasn't a robot. Tesla Bot was a dancer in a robot suit.

Literally. Musk dressed a person up in a robot suit and pretended it was a robot, in order to pretend to investors that his company would soon be able to grow by convincing manual laborers' bosses to fire them and replace them with robots, and split their wages with Musk.

Three years later, Musk demoed *another* humanoid robot—this time, a robot bartender that fluidly* mixed cocktails for the astonished audience.

This one *was* a robot, at least, but it wasn't an *autonomous* robot: rather, like the original Mechanical Turk, it relied on a tele-operator to control its movements in order to mix the cocktail.

Musk didn't even have the self-awareness to act

---

* Gettit?

embarrassed when these "robots" were revealed to be people pretending to be robots. And why should he? Once he convinces all our bosses to fire us and replace us with pretend robots, we'll all be so desperate for work that we'll sign up to pretend to be robots for pennies on the dollar.

Of course, it's *much* easier to pretend to be a chatbot than it is to pretend to be an ambulatory, humanoid robot. Many of the chatbot customer service reps, therapists, and other remote conversants are "backstopped" by low-waged workers in remote call centers who take over whenever the bot gets confused, without revealing that a human is now on the other end of the line. In spring 2025, Builder.ai collapsed: the company, once valued at more than $1 billion by its investors (including Microsoft), claimed that it had developed an AI system that would transform a customer's description of a new app into a functioning app. In reality, Builder.ai was a secret employment agency, farming out the work of building its customers' apps to eight hundred to one thousand low-waged Indian programmers. No wonder Indian wags have a new, wry joke for it: they say the "GPT" in "ChatGPT" stands for "Gujarati People Typing."

Agentic AI and Other Absurdities

. . . .

Displacing writers and artists makes for a killer demo, but writers and artists don't make enough money to make a dent in the hundreds of billions in capital expenditures hanging over the AI industry. To sell the story about how AI can recoup and realize a profit, the AI industry needs to spin a tale about replacing lots of waged labor, preferably *high*-waged labor. They've got a lot of stories about this, and none of them bear up to even cursory scrutiny.

Last year I spoke at an arts conference in Turin, Italy. This being Italy, they fed the speakers very well, and so the speakers' lunches and dinners were one of the highlights of the event. Plus, you got to meet the other speakers, who were a generally interesting bunch.

One day, I found myself sitting next to a pair of architects who had been invited to speak about how they were using AI in their practice. They talked around and around the subject for some time, but I kept pressing them for details. Eventually, they admitted that they were using AI to replace the draftsman whom they used to retain in order to create elevations based on the briefings they got from their clients.

The clients loved it—loved to sit next to the architects and give them feedback on their prompts and see the picture of their imaginary building change in response to their input, with just a few minutes' delay to see a new version. I get that—it's a lot easier to judge how a proposed change to a building will look if you can see a picture of it.

But much of the clients' pleasure is necessarily derived from *novelty*, from the sense of an intender behind each virtual brushstroke, which creates the impression that something *communicative* and *artistic* is being created at superhuman speeds, which can be a very heady sensation. Architectural elevations drawn by draftsmen in consultation with an architect are intended to communicate, not merely to illustrate. Eventually, the clients will figure out that they can sit at home, *not* paying an architect, getting an image generator to create pictures of buildings for them.

The architects at lunch (exactly what you're imagining: alarmingly bald guys with accents and round glasses) admitted that this was coming. But they were undaunted, because they were planning to extend their AI to go from an elevation to a blueprint.

Now, this *would* be an amazing trick. It's also not

something that an AI that makes pictures by filling in statistically probable adjacent pixels has any capacity to do. There is no meaningful relationship between a program that can make guesses about pixels and a program that can design a building. The former is guessing about how a building will *look*, without any understanding of how that connects to the structural elements a building can *contain*. A camera can faithfully record your appearance but does not (and cannot) do so by inferring things about your skeleton and internal organs.

The fact that you can write a program to draw a picture of a building by predicting the pixels that go next to the previous pixels, and so on, and so on, does not imply that the program can create a schematic for the building itself.

But it's not that hard to convince people that a program that can draw a picture of a building can draw its blueprint, too. Most of us don't understand how blueprints are made, and most of us don't understand how software works, and the Venn diagram of the two groups is more of a sphincter. That's a dynamic that is ready-made for creating an investment bubble.

We need to talk about bubbles, and then we'll get back to those poor artists.

Tech bubbles are surprisingly easy to generate, thanks to something economists call "the Byzantine premium." That's the extra value that investors place on an asset that they don't understand. They assume that any pile of shit of sufficient size *must* have a pony under it somewhere.

Many of the stories about AI's explosive growth potential have relied on this Byzantine premium, finding ways to exploit people's vague knowledge of esoteric subjects to impress them with decidedly unimpressive feats.

Take Google's proud announcement that its Deep-Mind AI had discovered so many exotic new materials that it had advanced the field of material science by *eight hundred years*. Then, some independent material scientists took a representative sample—380,000 of the 2.2 million "new materials" DeepMind had allegedly discovered—and found that *zero* of these materials were "credible," "useful," or "novel." They were either well understood existing materials (albeit with minor, irrelevant variations), or useless—for example, because they could exist only at absolute zero.

This is an age-old tech hustle: *This doodad can do something, so it will probably also be able to do something similar.* When I was a baby writer freelancing for *Wired*, my editor Bob Parks wrote a great investigative piece about a "child safety" anti-kidnapping bracelet (this was the 1990s, when stranger danger and child kidnapping paranoia were at a high not attained again until the QAnon years). The company that made this bracelet, GlobalTrak, had fitted it with a GPS receiver that could identify your child's location with a high degree of accuracy. GlobalTrak's pitch to parents was that if your kid gets snatched, this satellite communications device would use the GPS system to locate your kid and transmit their location back to you (and the cops).

Now, it may be hard to spot the sleight of hand in this pitch if you're a person living in 2025, where we have ubiquitous high-speed wireless data networks blanketing the world. But that's not what GlobalTrak was pitching: rather, they were promising to use the GPS system to transmit your stolen child's location.

There are *lots* of problems with this. For one thing, the GPS system—a constellation of thirty-two satellites that bombard the Earth's surface with highly precise time code that can be used to derive one's location—has no facility to *receive* data from GPS users. The idea that your kid's watch will use GPS to figure out where they are and to tell you about it is like claiming that your AM radio will deliver the morning traffic report and let you communicate back to the radio station about traffic jams they haven't spotted yet.

The other problem is that *sending* data to satellites requires large, powerful radios that consume a hell of a lot of battery and don't fit neatly into a wristwatch-sized package.

Now, surely the people who were marketing these crapgadgets knew all of this, but they were counting on us being such rubes that we wouldn't figure it out. They were asking us to make a logical—but fallacious—inference: that a device that could receive satellite signals could also *send* them.

It's natural to think that a program that can draw and redraw the facades of buildings on demand would also be capable of drawing blueprints for the building it imagines, but these are *unrelated capabilities*. These architects were either on the grift, or they were marks for someone else's grift.

Scratch stories about the significant market opportunities for AI tools with flashy demos, and you'll inevitably find a promise of AIs spontaneously developing new capabilities—capabilities that are not "doing the same thing they're doing now, but better" but rather "doing something no one has any idea how to automate using AI."

As I wrote this book in the summer of 2025, the world was swept by a wave of hype for "agentic AI," which is a marketing term for an AI to which you can assign tasks ("pick a place I'd like to go on holiday and find me a cheap plane ticket"), which it will perform with minimal or no supervision. Like "superintelligence," "agentic AI" is a new source of inflation for the AI bubble, something that tech bosses can sub in for last year's "generative AI," which is getting tired as the novelty wears off and people discover that "gen AI" mostly does parlor tricks, and can't substitute for the valuable labor we were promised it could displace, justifying its massive investment.

The first generations of AI agents were *terrible*. They'd load a web page, take a screenshot of it, then try to reason out which regions of the screen they should fire a click at to get at the part of the website they were trying to access. This worked about as badly as you would imagine, and only got a little better when the developers started tasking the AIs with looking at the HTML and Javascript source code for web pages as a way to figure out which parts of them to click.

When this approach failed to yield fruit, the agentic AI industry announced a future of *specialized* agents, each of which would be trained to understand and manipulate a range of websites and services in a narrow field of focus. Users could daisy-chain these agentic AIs together, say, by using one agent to assess different vacation destinations, another to suss out travel arrangements, a third to figure out travel visas, a fourth to book tours and events, and so on.

These specialized agents still have to navigate a complex world of heterogeneous websites. This is challenging, and not just because these sites weren't designed to be legible to AI systems, but also because they were designed to frustrate *humans* seeking to gain an advantage by understanding all the contours of the offers they presented.

The Biden administration's Federal Trade Commission spent years doing pitched battle with websites (especially travel websites) over the proliferation of "junk fees"—the nonsensical surcharges that show up on your screen once you've spent a long time agonizing over the details of your purchase. Merchants use junk fees to prevent apples-to-apples comparisons between their offerings and those of their rivals, displaying prices that can't be readily compared, because the *actual* price isn't revealed until you've performed a gigantic amount of tedious, fiddly labor to specify and complete your purchase. Companies don't want you to get other useful information about why you might prefer a rival's product over their own, because the less information you have, the more they can fool you into buying things that are more expensive and/or of lower quality.

Companies go to enormous technological and legal lengths to prevent third parties from indexing their pricing information and other information that is critical to comparing their offerings to rivals'.

They even use AI to reprice their offerings based on predictions about how much you're willing to pay (or accept). As previously discussed, for example, in July 2025, Delta Air Lines announced a partnership with an Israeli AI pricing startup called Fetcherr, founded by former hedge fundies. Peter Carter, Delta's CEO, told investors that Fetcherr's AI tool would be able to consult the vast troves of surveillance information that data brokers assemble on every American before showing fliers a price for a plane ticket, so that they could charge more based on the flier's predicted willingness to pay.

After public outcry, Delta insisted that their CEO's absolutely unambiguous statements were being misinterpreted. It's possible that Peter Carter was lying in order to impress his shareholders by "integrating AI" with his mature legacy business, in hopes that they would revalue it as though it were a buzzy tech company. Whether or not Delta is using AI and surveillance to reprice its goods, there are certainly many industries that are doing "surveillance pricing," as an investigation by Biden's FTC showed.

(The Trump administration allowed these junk fee and surveillance pricing efforts to die.)

Surveillance pricing can be applied to buying labor just as readily as it can be applied to selling goods and services. Uber uses data about its drivers—specifically, whether they take all the rides the algorithm throws at them—as a proxy to gauge how economically desperate they are, and lowers the wages of drivers who are inferred to be in dire financial straits. Veena Dubal calls this "algorithmic wage discrimination," and while Uber invented the practice, they've been overtaken by firms from other industries.

Surveillance pricing is one example of how firms use "information asymmetries" to rig prices to the detriment of customers and workers. More broadly, the fact that digital services allow for continuous, algorithmic repricing (including algorithmic repricing based on the prices charged by direct competitors in highly concentrated sectors) means that firms can increase their profits by obscuring their prices from scrutiny.

And they do! Not only do companies use junk fees and individualized pricing to prevent comparisons and analysis, they also employ engineers who design and maintain elaborate anti-scraping measures. They also sue and threaten people and companies who figure out how to scrape and analyze the data they publish about their prices and offerings, whether that's for people selling their labor (DoorDash shut down the Para app, which allowed "dashers" to see the full wage, including tip, before agreeing to take a job), or buying goods and services (American Airlines and Ryanair have both attacked services that scrape and analyze their prices).

All of that makes agentic AI even more improbable. Companies don't *want* software agents to automate away the high barriers they've created to prevent direct comparisons and analysis of pricing cycles. They will use the law *and* technological countermeasures to prevent this.

Which makes the next phase of the agentic AI pitch even more stupid. After agentic AI failed to live up to its promise of a single agent that could navigate the designed-for-humans web to automate complex tasks, and after specialized agents tailored to subsets of the designed-for-humans web failed to live up to their promise, agentic AI pitchmen announced that they could make the whole system work by creating robust public standards by which websites could reveal themselves to AI agents in order to facilitate agents' retrieval of their information.

This is a *great* idea. Since the late 1990s, advocates for the "semantic web" have been trying to convince website operators to create "structured data" that hewed to standard "vocabularies" and "ontologies" so that software agents could easily locate and analyze the data they published.

This has not happened.

It hasn't happened for a lot of good reasons. For one thing, there are irreconcilable differences in the ways in which things should be described. A site run by transphobes isn't going to label every story about a trans woman as "a story about a woman," irrespective of what the standard says. Microsoft is never going to label Excel as equivalent to Apple's Numbers, Google Sheets, and LibreOffice's Calc. Hermès is never going to list a $40,000 handbag as being made of the same material and having the same capacity as a $30 leather handbag of the same size and shape.

It also hasn't happened because there is a reward for cheating. Scammers will label their phone numbers as the customer service number for your cable operator or bank. Counterfeiters will label their batteries or phone cases as official accessories. It's hard enough to get people to correctly label their stuff—but *fact-checking* billions of labels created by millions of people, none of whom you know personally or whose identities you can verify? That's *insurmountable*.

Finally, it hasn't happened because *companies don't want you to have good information about their products and services, their prices and costs, because when they know things you don't, they can make more money off of you.*

The semantic web was always a political nonstarter for these reasons, but at least it was created as a collaborative, open project that stood to benefit many different kinds of stakeholders.

By contrast, the agentic AI industry is now insisting that everyone rearrange all of their affairs to the benefit of AI companies, and to their own detriment. This is an extremely unlikely (and unreasonable) proposition. It's a sleight of hand: *We can do amazing things with this technology (once we get everyone else in the world to change everything about how they work, to their detriment and our benefit).* From stock analysts to tech journalists, this message is everywhere, from the pages of *CIO* magazine and *The New Stack* to investment advice from AI-touting financiers.

This isn't even the first time that the AI industry has pulled this trick. Many companies raised tens of billions of dollars by promising that they would soon field "full self-driving" cars. This is the promise underpinning the sky-high valuations of companies like Uber and Tesla.

While "full self-driving" cars have made some progress in recent years (see, for example, the Waymos on the streets of San Francisco, Phoenix, and Los Angeles, and the Tesla Robotaxis in Austin), these cars continue to experience worrying lapses—driving on the wrong side of the road, say. What's more, there's no information on how these cars drive under *adversarial* conditions: earlier iterations of Tesla's autopilot could be tricked into driving into oncoming traffic with strategically placed pieces of tape that made its computer vision system go berserk. It can feel pretty cool to ride around in a robot taxi that pilots itself eerily through the streets, but these things are not mainstream, nor are they close to being mainstream.

A whole field of machine learning research called "adversarial examples" shows how the "deep learning" model employed by the systems in the latest AI bubble—from LLMs to vision systems—are vulnerable to this sort of attack. In 2023, prankster protesters in San Francisco showed that you could immobilize a GM Cruise robotaxi by putting an orange traffic cone on its hood. The car's autopilot would interpret this as a traffic cone directly in its path, no matter which way it turned, and since it was programmed not to run over traffic cones, it would give up and sulk.

Cruise robotaxis no longer patrol the streets of San Francisco, or anywhere else. That ended in 2024, after a Cruise vehicle hit a pedestrian and dragged her twenty feet before stopping. During the ensuing investigation, Cruise was forced to admit that every one of its vehicles was overseen by 1.5 human engineers, who cost more by the hour than any taxi driver, especially now that Uber and Lyft have eroded drivers' wages to sub-starvation levels.

Other robotaxi companies (notably Waymo, part of Google) have fared better than Cruise, and have been able to charge a premium for the privilege of riding in a car that either has zero drivers or an unknown number of drivers remotely operating and monitoring it (and you, as you ride in it—Waymo isn't saying). If Waymo wipes out the wage bill of every rideshare driver, they will not recoup the extraordinary cost of building this technology. But for Google, the loss might be worth it, if the company can maintain the impression among investors that it is still growing.

For Cruise—and its owner, General Motors—replacing one low-waged driver with one and a half high-paid engineers had been a gamble worth taking, because it was a gamble that investors would decide that a tired old "mature" car company with few growth prospects was actually a *tech* company with incredible growth potential that would justify an overnight jump in the company's price-to-earnings ratio, which might let GM hire so many engineers and acquire so many self-driving car startups that they actually *become* a tech company. And if not, well, lots of GM insiders would be able to sell their stock at a historic high.

Cruise's horrifying robot-on-human assault spurred a fresh wave of public discussion of the safety of robo-cars. Andrew Ng, an advocate (that is, shill) for self-driving cars, told *The Verge*'s Russell Brandom:

> I think many [autonomous vehicle] teams could handle a pogo stick user in [a] pedestrian crosswalk,

having said that, bouncing on a pogo stick in the middle of a highway would be really dangerous.

Rather than building AI to solve the pogo stick problem, we should partner with the government to ask people to be lawful and considerate. Safety isn't just about the quality of the AI technology.

This is the same move that the agentic AI people are pulling today: if a self-driving car hits you, it's because you weren't "lawful and considerate." Self-driving cars can drive themselves, sure, but only if we change the behavior of every other road user, and also the law of the road, so we can punish those "lawless and inconsiderate" people who do something the AI couldn't anticipate.

Indeed, the auto industry *invented* this gimmick when they answered widespread outrage about people being killed by rich idiots in cars (when cars were toys for the wealthy) by inventing the idea of the "pedestrian," who shouldn't be in the roadway, where people had walked since roads were invented. The idea of a "pedestrian" (and that other invention, the "jaywalker") shifted the blame for product-related fatalities away from the manufacturer and its customers and onto the people who were mangled under the wheels. Today, the robotaxi people want to blame pedestrian injuries and deaths on the "pogo stick problem" and agentic AI misfires on badly designed websites. The world is both unpredictable and adversarial, and any "autonomous" system that works only when everyone co-operates with it does not work at all.

For AI to be anything other than a bubble, the world will have to rearrange itself, at farcical expense, to make life easier for chatbots and their cousins. When this doesn't happen, the bubble will burst.

AFTER

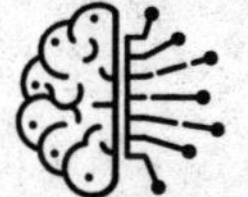

Bubbles and Productive Residues

.....

AI is a bubble, but not all bubbles are created equal. Start here, though: every bubble is bad. The foundation of a bubble isn't just "irrational exuberance" in which retail investors—you and me—unwisely gamble our life savings and lose everything. Bubbles all have *winners*, and these winners don't come out on top by accident. Before a stock swindler inflates a bubble, they first acquire a lot of whatever the bubble is *for*, so they can sell it to us, and make out like a literal and figurative bandit.

Every bubble is a transfer of wealth from savers to crooks. Every bubble is bad. We shouldn't have bubbles. The pump-and-dumpers who inflate these bubbles should face criminal sanctions. Regulators should intervene to prevent bubble formation in the first place.

That said: not all bubbles are created equal.

Some bubbles pop and leave nothing behind. These are the pure fraud bubbles.

Think of Enron: the company made billions by claiming to have realized new efficiencies through "energy trading" in which access to electricity was treated as a financial instrument and swapped back and forth by gobby traders who ripped off everyday people while imposing rolling blackouts on millions of innocents. When Enron popped—and its C-suite were sentenced to prison—there was nothing left over, save some greasy boiler-room telephones and used servers full of victims' personal information and the private emails of Enron employees.

Or take cryptocurrency. Crypto isn't money: it's too volatile to be a store of value or unit of account, and it's too cumbersome to be a medium of exchange. Crypto is a form of pure speculation, a way to bet on whether someone else will think that the "coin" you've just bought is worth more than you paid for it. This is a "greater fool" swindle, in which fools buy something in the hopes that a greater fool will buy it from them at a profit.

The first crypto bubble was *wild*. At one point, Long Island Iced Tea—makers of a failing brand of sugary, flat soft drinks—rebranded itself as the "Long Blockchain Corp." and saw a *380 percent* jump in its valuation, despite having nothing to do with blockchains or cryptocurrencies apart from changing its name.

The crypto bubble keeps on getting reinflated, not least because the literal president of the United States issued his own shitcoin, putting a figurative tip jar right there on the *Resolute* Desk in the Oval Office. But eventually, the crypto bubble will burst (again) (and permanently) and when it does, what will be left?

Well, there'll be a smattering of programmers who've learned something about cryptography (a very important subject that we can always use more experts in, civilizationally speaking), and many of them will have learned good coding habits using new, secure programming environments like Rust, a programming language that's widely used in cryptocurrency projects.

But apart from that, when the crypto bubble bursts, the only thing it will leave behind is bad monkey JPEGs and worse Austrian economics.

But many bubbles *are* productive, even if they are dishonest and destructive at the same time. Take the dot-com bubble of the late 1990s and early 2000s: while that bubble did drain hundreds of millions of dollars from pension funds and tank the life savings of individual savers, it *did* leave behind more than foosball tables, lightly used Aeron chairs, and promotional T-shirts.

The dot-com bubble funded the entry of millions of humanities undergrads and misfits into the tech sector, subsidizing their instruction in HTML, JavaScript, Perl, and Python, and all that expertise hung around after the bubble popped. It's where we got the best of Web 2.0, as people from nontechnical backgrounds brought a radically different perspective to the project of online service design.

That wasn't the most efficient or honest way to get new

blood into the tech sector, but at least we got *something* out of it.

Then there's WorldCom, the wildly fraudulent enterprise whose CEO, Bernard Ebbers, died in prison after his part in a global-scale stock swindle that manufactured fraudulent evidence of insatiable demand for fiber-optic connections. WorldCom sucked up millions in investor capital under false pretenses—and inspired many copycats that raised money on the assumption that WorldCom wasn't a giant fraud.

WorldCom was a giant scam, but they did lay a *hell* of a lot of "dark fiber" all over the world. Fiber basically lasts forever (it's spun glass), which means that under the streets of cities everywhere, there is WorldCom fiber, waiting to be connected to routers at a data center and lit up. My old street in East London had steel utility covers set into the sidewalk, stamped with the WorldCom logo. And at my current home in Burbank, California, I've got two-gigabit symmetrical fiber through AT&T, which was able to buy abandoned WorldCom fiber at pennies on the dollar.

WorldCom was a scam, its CEO was a sociopath, but it left behind a productive residue.

So what kind of bubble is the AI bubble?

AI is incredibly capital-intensive to create. Billions of files have to be assembled and hand-"tagged" by human labelers who describe the training data so the AI can know how to slot it into its statistical model.

Then, this data has to be subjected to incredibly computationally intensive analysis, consuming enough energy and water (used to cool the computers performing the analysis) to satisfy whole cities. The environmental costs of this "compute" is off the charts. Even if you stipulate that the world will benefit from having some giant, "advanced" AI tools, there's no rational case for endangering the planet and the lives of millions of people to make several redundant AI tools that are functionally indistinguishable, with each consuming so much energy that they wipe out a substantial share of the gains made from solarization and the broader switch to renewables.

For companies hoping to demonstrate their superiority to Wall Street, using *lots* of energy is a way to send a signal about how "advanced" their technology is. Energy consumption is a rough but easily understood proxy for computation, and if we naively assume that all computation is efficient and necessary, then the more energy a company consumes, the more necessary, efficient computation it must be doing. As I wrote earlier, investors are like the kid who exclaims with joy and jumps straight into a pile of manure, shouting that "A pile of shit this big *must* have a pony under it *somewhere*!"

This kind of computation is very rough on computing hardware, with the result that the GPUs—the specialized "graphics processing units" that AI data centers are stuffed with—burn out at an incredibly high rate. During a single fifty-three-day training run for Meta's Llama model, the company found that *more than half* of its GPUs burned out. These GPUs, like all microprocessors, are stuffed full of conflict minerals, and their production consumes vast amounts of energy and water and results in the emission of rivers of toxic sludge.

Then, once these foundation models are trained and "tuned" (tweaked by both skilled and semi-skilled testers and technicians), they need to run on more giant, over-taxed GPUs, consuming more water and energy—and contributing to more carbon emissions—every time they are accessed by a user.

When we hear about the energy consumption and climate impacts of AI, we're hearing about these foundation models: expensive and wasteful to create, and expensive and wasteful to operate.

This is a trend that is only getting worse.

Defenders of AI will often cite other technologies that were costly at the outset, like the web itself, as evidence that AI will soon solve all its energy and scale problems. But the web revolution was a decade of unbroken drops in the cost of servicing each new web user. In economics terms, the web had *great* "unit economics"—the cost of each web session fell and fell, even as the cost of getting on the web (from hardware to software to connectivity) was *also* falling.

Not so with AI. Each generation of AI foundation models has been *vastly* more expensive to train and operate than the previous generation. Not only that, but many refinements in AI that are meant to improve accuracy and reduce "hallucinations" involve breaking a prompt down into multiple pieces and prompting an AI to respond to each prompt, turning that response into a new prompt, over and over again, to produce "chains of thought."

These are, to quote Ed Zitron, "dogshit unit economics." AI gets more expensive every time it adds a user. It gets more expensive every time it adds a feature. It gets more expensive every time it improves. This is the opposite of the conditions under which the web attained liftoff.

Companies are *not* training new foundation models because the old ones are so profitable that their owners have gobs of cash left over to make even better ones. Every AI company is losing money, and the bigger the AI company is, the more money it's losing. The biggest AI companies are losing *billions* per *quarter*.

The financing for new models is coming from investors, and those investors are making a bet that the AI they're funding will be so good that employers can fire half their workers and replace them with AI, with the proceeds split between AI companies' customers, and the AI companies themselves. Failing that, they're making a bet that AI companies' sales staff can convince employers to fire half their employees and replace them with AI that *can't* do their jobs, and that no one will figure this out until after the AI companies' investors have cashed out.

Given that every major improvement in AI to date has been vastly more expensive than previous ones to launch, and that the only companies who claim to be saving lots of wages thanks to AI are companies selling AI products, this bet is a *long shot*.

What's more, even if it turns out that *an* AI company manages to be profitable someday (a big *if*), this does *not* mean that the rest of them will succeed. Indeed, given the extremely high costs associated with creating and operating a foundation model, it may be that foundation models are "natural monopolies" (products or services that can support only a single seller). I'm sure a lot of AI investors believe this, and consider it a feature, not a bug, since they think *they're* backing the AI company that will emerge victorious and claim the *whole* market for AI services.

Even if one AI company survives, most or all of the rest of them will fail. When that happens, the AI bubble will burst. Investors will panic-sell their holdings. Companies will collapse.

In other words, the bubble will burst, as all bubbles do, always.

No matter how much you hate AI, this will not be a good day. Remember: seven giant AI companies account for 35 percent of the U.S. stock market. Amputating 35 percent of the market is going to *destroy* a ton of innocent bystanders, including people whose retirement savings are invested in index funds, considered the safest of all safe bets. We're talking about a crash that will put 2008 in the shade and meet or exceed the pandemic sell-off.

The great economist John Kenneth Galbraith had

a name for the time before the collapse. He called it a "bezzle," which he defined as "the magic interval when a confidence trickster knows he has the money he has appropriated but the victim does not yet understand that he has lost it." Right up to the moment when the AI bubble pops, there will be bosses who think they've found an amazing, reliable way to slash their labor costs. There will be investors who'll think that they've bet on a sure thing. There will be everyday people who'll think that they can rely upon their AI romantic partners, email writing tools, and vibe-coding assistants.

After the bubble pops, though, most of that will disappear, forever, and a few hustlers who got out early will make out like bandits.

It's possible that the world's governments will step in to bail out the sector, but really, how much can they do? Every day you leave an AI data center running is a day you lose money. The more popular your AI service is, the more money you lose. What's more, that AI center is literally incinerating million-dollar GPUs *all the time*, and these have to be constantly replaced. A onetime bailout isn't going to keep the foundation models' lights on.

The finance sector loves a grift, but as Stein's Law has it, "If something cannot go on forever, it will stop."

So what kind of residue will this bubble leave behind when it pops?

One obvious remnant will be the physical plant and its associated hardware: all those data centers and GPUs.

There are lots of scientific applications for massively parallel computers (ironically, climate modeling is totally dependent on this kind of infrastructure). What's more, you don't *have* to churn through GPUs at the rate of naughts—that's down to AI companies sweating their capital like it's a government mule.

There will also be a lot of people who will have been trained to program and refine AI-type statistical engines, even more people who'll have a ton of expertise in prompting AI models, and even more people who'll be good at labeling data so that it can be retrieved and manipulated by computer programs.

It's not clear how well these skills will serve these workers after the bubble pops and the unlimited fire hose of capital gets switched off. But if you want to start planning for some kind of project in the future, you could do worse than taking it as a given that you'll have discount access to a lot of parallel computation, applied statisticians, data labelers, and "prompt engineers."

This isn't nothing. Maybe in fifteen years we'll all marvel at the massive breakthroughs made in, say, weather simulation, when all those GPUs hit the bankruptcy auctions.

Realistically, though, very few of us are going to start companies or nonprofits or hobbyist projects that reuse the capital from the AI bubble or hire its laid-off staff.

On the other hand, there is a kind of AI residue that I believe *many* of us will continue to use, long after the AI bubble has popped: the stand-alone, open-source, "toy" models that run on commodity personal computers and devices that are even cheaper and lower-powered.

I started this book with an anecdote about how I was able to download thirty hours of podcasts and run them through Whisper, a free, open-weight model that OpenAI irrevocably released, in order to transcribe them. Whisper did a *fantastic* job, took a remarkably short amount of time, and consumed so little compute that my laptop's fan didn't even switch on.

Thanks to models like Whisper, we are likely to all have extremely high-quality transcription software on our computers that we can just throw any recorded speech at in order to search it. Future operating systems might even have a background process that takes advantage of idle processor cycles to transcribe *all* the video and audio on your computer, so you can search it the same way you might search the contents of your email archives.

On-device image processing is also incredibly impressive. There's a lot of attention (rightly) paid to how bad people use open-source image generators to produce nonconsensual deepfake nudes or media hoaxes on their home computers, but there are so many ways that image processing can function on-device.

For example, my phone—a two-year-old Google Pixel—can use on-device processing to transcribe all the text in the photos I take. More impressively, the on-device image-processing software that's bundled with the phone can remove people from the background of photos, making shrewd guesses about what's behind them. I take a lot of photos and upload them under Creative Commons licenses that give anyone the right to use them as illustrations for their own projects, and I try not to include strangers in these photos, because I don't assume they consent to being stock art. A smart airbrush that can remove an incidental face or two makes those photos much more useful for everyone.

On-device image generation is also extremely useful. Thinking about a kitchen remodel? Find some inspiring samples and ask the image generator to create a bunch of combinations to discuss with your contractor.

As a collagist who can't draw *at all* and limits himself to public-domain and open-access sources, I've been thwarted many times because I needed a graphic element that just wasn't available—say, a picture of a horse's head from behind (perhaps to matte onto the body of a carefully cropped public-domain image of a coal miner, to create a reverse centaur). Having a local image-gen model on my laptop that can whip up those perfect little image elements would be fantastic.

There are plenty of very *fun* uses for this stuff, of course. I have friends who play a weekly Zoom game of Dungeons & Dragons. The dungeon master uses a local model to transcribe the game and summarize it, and another local model to produce images of key scenes in the game, all of which are distributed to the players after game night. The (many) errors the AI makes in generating these post-game summaries are funny *because* they're so miscon-strued (like the early auto-translate that converted "out of sight, out of mind" to "invisible idiot"). The extra fingers and other oddities in the pictures are likewise funny and charming in this context. The image generator is work-ing from a massive corpus of transcribed play to use as its prompts, which makes for a very communications-dense form of prompting, enlivening the images.

This may seem like a paltry and trivial outcome from hundreds of billions of dollars and gigatons of planet-killing carbon, *because it is*. The serious environmental costs of AI will be borne by everyone on the planet, and all the animals and plants we share it with, for hundreds of years to come. Much of that harm is already locked in, and all we can do now is think about how we will miti-gate it—how we'll treat the zoonotic plagues, where we'll house the climate refugees, how we'll evacuate our low-lying cities, and how we'll douse the wildfires.

The economic costs of the AI crash will also be felt around the world, for a generation or more. Those harms are *also* locked in. You can't give a third of the S&P 500's value over to seven money-losing AI companies that en-ergetically pass the same $100 billion IOU around and around without creating the conditions for a prolonged, brutal, global crash.

Remember: all bubbles are terrible, but some bubbles are productive. The stuff that's left behind when the bub-ble pops is *salvage*. Just because we deplore the waste that went into its production, that doesn't mean we have to contribute to that waste by discarding this perfectly good remnant.

There are plenty of uses for chatbots, too. Many, many people report that "conversing" with a chatbot is therapeutic. That's easy to believe. After all, the very first chatbot, ELIZA, created in the mid-1960s by the computer science and visionary technology critic Joseph Weizenbaum, was a simulated therapist. ELIZA was very crude by modern LLM standards, but people reported that "conversing" with ELIZA was often revelatory.

A chatbot "therapist" is more like an interactive journal, one that feeds you back bland—but increasingly personalized—responses as you spill your guts to it.

To put this in the context of AI art: AI art is uncanny because it has the seeming of intent without an intender, and it grows more meaningful the more a human infuses it with communicative intent. An AI therapist is a chatbot that you iteratively prompt and re-prompt, and the sentences in its replies to you are a higher- and higher-fidelity expression of the communicative intent you infuse into the sentences you feed it.

An AI therapy session is a creative work of literature in which you are the principal author and the chatbot is a coauthor that makes successively refined guesses about what you want to hear based on your responses to its responses (to your responses to its responses).

Some people have experienced "AI psychosis" in which

they have led their chatbots into extremely destructive conversations wherein the chatbot fed and amplified dangerous delusions. At least one person killed themselves after such an experience.

These are extremely tragic occurrences and they warrant research into how to create chatbots that can brake their users' descent into self-destructive spirals. But there's a fundamental difference between your choice to talk to an AI with your computer, and a for-profit company putting its robotaxis onto public streets and conscripting everyone around them into participating in a live-fire test of its ability to safely navigate the world.

But for the majority of people who find a chatbot to be a willing, friendly ear and crying shoulder, the existence of stand-alone models that run on their own computers is a huge boon. These chatbots deliver all the value of an AI therapy system without any of the costs—not the monetary or human costs. Many "AI therapists" are backstopped by a human in the loop—a human therapist who monitors a screen full of therapy chat sessions, hopping from one to the next as the AI flags human inputs for a supervisor's attention. This is not "therapy," and it's a terrible job that crushes anyone who cares about helping people. No one should have this job. With a stand-alone model running on your own computer, the job ceases to exist.

In addition to all those costs, a local chatbot that acts as a kind of therapeutic, interactive journal does not incur any of the *privacy* costs associated with AI therapy services.

AI companies are the most data-hungry, privacy-disrespecting firms in tech history, which is a tough, highly competitive field. They *should not be trusted* with the compromising, sensitive personal reflections of their users. They have never found a data source they will not scruple to ingest, and highly personal therapeutic discussions are a kind of training material that is hard to find in the wild. The chance that one or more of the AI companies will be found to have trained a bot on therapeutic sessions is 100 percent. The chance that even more will do so, but *won't* get caught, is also 100 percent. Anyone who trusts an AI therapist run by an AI company needs their head examined.

But the point is that stand-alone open-source models that you run on your own computer mean that you don't *have* to trust an AI company with your therapeutic self-talk, if you find that kind of interactive journaling helpful.

The AI companies that spun out open-source models clearly intended these to serve as advertisements for their subscription-based, cloud-hosted foundation models (additionally, Meta probably hoped to convince a bunch of software developers to devote their time to mastering PyTorch, Meta's AI-centric programming tool). These companies were in for a surprise.

If the cloud-hosted foundation models have constantly *underperformed* relative to the boasts of AI companies, their stand-alone open-source models have *overdelivered* relative to the companies' stated expectations. Communities of open-source developers and model users have pushed and pushed and *pushed* the boundaries of what these "toy" models can do while running on your laptop or phone, like hardware hacker Pete Warden, who's demoed a voice-prompted, talking manual for small appliances that runs on a chip so cheap and low-powered that it costs less than $10. This model can understand your questions about troubleshooting a dishwasher, locate the correct trouble-shooting instructions in a stored manual, and walk you through them—on hardware so cheap it could come with a singing greeting card off a spinner rack at a Walgreens.

Even more so are the commercial operators who took these open-source models and *ran* with them. No model exemplifies this more than DeepSeek, the Chinese model that was created as a side hustle by High-Flyer, a hedge fund.

Constrained by a U.S. blockade that limits Chinese companies' access to the most powerful GPUs, High-Flyer refactored the standard techniques for running a model to make it run more efficiently on cheaper, lower-powered chips. The results were *stunning*, delivering a level of performance comparable to the best cloud-based foundation model with open-source code that could be run on commodity hardware.

The release of DeepSeek in 2025 sent shock waves through AI investors. DeepSeek laid bare the incredible *laziness* of the giant U.S. AI companies, who solved all their scaling issues by throwing money at their problems, rather than by applying their ingenuity to them. Of course, if you are less interested in making useful AI prod-ucts than you are in convincing investors that you're do-ing something with enormous growth potential, it makes sense to spend as much money as possible. Investors may not understand how AI code works, but they *do* under-stand that things that are *very* expensive are difficult to replicate and thus the more an AI model costs to produce, the fewer competitors it is likely to have.

DeepSeek's debut sent a cold chill up the spine of ev-ery investor in a big U.S. AI company. If their $100 billion models can be bested by a model that cost a reported $6 million to create and can run on commodity hardware, what future do these top-heavy AIs have? It's pretty fun to watch someone build the world's heaviest airplane, and it's pretty cool to see it fly, but that thing isn't going to be able to compete with an Airbus.

The sell-off that DeepSeek prompted was *spectacu-lar*. Nvidia's valuation alone fell by two-thirds of a *trillion* dollars, in *one day*. Remember: a growth stock has a very high price-to-earnings ratio; a mature stock has a P/E ratio that's a mere fraction of an otherwise similar com-pany that has plenty of growth potential. That means that once the market concludes that a firm has transitioned from "growth" to "mature," its stock experiences a sudden and precipitous decline. Anyone still holding that stock after the crash will have lost their shirt, so the *instant* a company's growth seems to be petering out, investors panic-sell their stock to get out ahead of everyone else.

Nvidia's share price came back up—this time. But it won't keep bouncing back forever. Anything that can't go on forever eventually stops.

While some open-source hackers have pushed the limits of those models to do awful things, the vast majority of the achievements from the model-hacking community have been brilliant and delightful. These communities typify the ingenuity, playfulness, and resourcefulness of the old, good internet. This has got to be *terrifying* to people hoping to get rich by making hundred-billion-dollar AI models, but even better—these open-source hackers are going to be around long after the bubble that funded those giant, bloated, destructive models has burst and all that's left of those AI companies are bankruptcy auction GPUs, foosball tables, and enough defunct company logo shirts to supply every brand necrophiliac alive today or yet to be born.

If we are good AI critics, if we carefully identify the pathological aspects of AI and relentlessly target the financial basis of the AI bubble, we can make sure that all the terrible things billionaires want to do with AI never happen, and we can consign all the terrible things that are *currently* being done with AI to history's ash heap.

AI After the Bubble

. . . .

After the AI bubble pops, we can finally get down to the business of assessing what these applied statistics tools can do for us.

Because they *can* do things for us. The centaurs who are pleased and even amazed by the use they get out of their AI tools aren't (necessarily) deluded. There're plenty of real, beneficial use cases for AI, and even better, many of these use cases can be executed on stand-alone open-source models that run on commodity hardware—that is, on models that we'll continue to have access to for as long as we like, bubble or no bubble, simply by installing some software on our computers.

As a lifelong activist with the Electronic Frontier Foundation, I've been lucky enough to be exposed to AI uses that have tremendous utility in the struggle for human rights.

Earlier in this book, I introduced you to the Human Rights Data Analysis Group (HRDAG), founded by the mathematician Dr. Patrick Ball. HRDAG is a real standard-bearer for the socially beneficial uses of AI. For decades, HRDAG has collected and analyzed data about genocides, war crimes, and mass-scale human rights abuses, using rigorous statistical methods and expert scientific communication skills to inform truth and reconciliation proceedings, war crimes tribunals, and other high-stakes forums.

HRDAG has found many ways to use AI to improve their work. Take data collection: HRDAG does extensive work on wrongful arrests, police violence, and police abuse in the United States (in 2015, they did the first-ever national census of police killings of civilians).

To obtain the data for this work, HRDAG works with organizations like the Community Law Enforcement Accountability Network (CLEAN) to file public records requests with police departments.

As HRDAG's Tarak Shah explained to me, "These requests yield dozens of terabytes of unstructured data: scanned documents, video footage, audio recordings of interviews, photographs of evidence, and more. The files arrive in a disorganized dump, without indication of what they contain or what they are about. We use LLMs to extract data from the files to link files together that are about the same investigation, as well as to build out indexes of people, places, and events that can be used to search the data. They use LLMs to help identify sensitive private information about victims and witnesses that agencies failed to redact, in order to publish records in a way that is responsible and does not perpetuate harm. The LLM-based extraction also allows it to monitor compliance: we are able to evaluate whether a disclosure about a case is complete or is missing a key document, we also compare LLM-extracted data from agency disclosures to available data in other public data sources in order to further identify records that should have been disclosed but were withheld. In several jurisdictions we were able to use these analyses to take agencies to court to compel fuller disclosures."

Pre-AI, these scans were *really hard* to turn back into data. But large language models are capable of making good guesses about ambiguous words, and, even better, turning scans of tabular data *back into tables* that can be directly manipulated with software. AI turns out to be *really* good at this, because tabular data often has a grand total at the bottom of each column, so if one cell is hard to make out (for example, if you can't tell if a number is a three or an eight), the AI can automatically check to see which of the possible numbers would produce an accurate total.

If this sounds trivial, that's because you're not thinking big enough. The ability to do this *at scale* allows HRDAG to build up *massive* databases of vital human rights statistics.

For example, HRDAG partnered with Innocence Project New Orleans to ingest mountains of arrest reports and flag ones that contained the correlates of false arrests (arresting officer, circumstance, etc.). This narrowed the field for Innocence Project workers, who investigated the flagged reports further and fed them into their exoneration work.

In California, HRDAG was able to ingest vast numbers of parole records and compare outcomes for similarly situated prisoners to quantify the exact degree to which California's parole boards practiced racial discrimination in granting parole.

In Colombia, HRDAG worked with the national truth and reconciliation process to analyze fifty years' worth of records of extrajudicial killings during the country's brutal civil war. They were able to use machine learning to "de-duplicate" these records (merging multiple records of the same killing that appeared in different datasets). Then used other machine learning systems to score each killing, estimating the likelihood that it had been committed by a CIA-backed right-wing militia, a leftist FARC guerrilla, the military, or someone unrelated to the civil war (for example, a killing in a bar fight).

This allowed HRDAG to present a picture to the Colombian people of how many of their number had been lost in the civil war, and to break down those killings by perpetrator, showing the largest and smallest number of killings that could be attributed to each party (spoiler: the CIA-backed militias were Colombia's most prolific murderers).

I raise HRDAG's work here because they are so well

situated to take advantage of the beneficial uses of AI. Small, cash-strapped, with a huge mission, HRDAG needs every inch of leverage it can get. And because HRDAG is so technically clued in—the scientists who work there could be earning ten times as much in industry in a heartbeat—they immediately saw the potential for these tools.

If these AI models had been developed in a corporate or university lab, trained and released into the wild or hosted on a big academic cloud, we'd all know a thousand stories like this one. But every story of an organization using AI the way HRDAG does is drowned out by a hundred breathlessly reported stories of some nonsense someone is getting up to with AI—nonsense that does no one any good and is going nowhere.

After the bubble pops, we're going to have a wide array of new and extremely useful tools at our disposal. Everyday image processing is going to get a *lot* better (and weirder, because inevitably, there will be some AI errors creeping into the red-eye, blur, and other correction filters in our phones). Most spoken word will be easy to transform into searchable text.

We'll have better translation tools, too: not tools that replace interpreters who do real-time translations for conference speakers and diplomats, nor the translators who prepare foreign editions of books and papers. But we will certainly get better tools for things like the raucous conference calls my extended family gets on with our relatives in the former Soviet Union, which are presently a hilarious babble (and Babel) of Yiddish, Russian, and English, along with machine-translated chat text. Likewise, travelers and refugees will have far better tools for navigating and communicating with people that they don't share a language with.

We'll doubtless find some fun ways to use these tools in games—not to wholesale replace the writers and actors who bring nonplayer characters to life, but to give those NPCs a more open-ended set of interactions and responses.

And, of course, we'll have all the fun uses—the D&D game that is transcribed, summarized, and illustrated (along with many comical failures) by a chatbot and an image generator.

If all this stuff turned up as new features for existing products—superior text-to-speech for your phone, a better Google Translate, cool new filters for your camera—we'd be pleasantly surprised and find lots of ways to improve our lives with it in ways small and large. Programmers would work the stuff into their coding workflow, treating it as just another tool that automates a few boring tasks and serves as an error-checking backup that highlights possible mistakes.

It's because this all arrived as part of an investment bubble that was inflated by attacking workers of all kinds, and that stands to destroy the planet and the economy, that we are forced to take sides as either "anti-AI" or "pro-AI."

It's fine to be "anti–AI *bubble*," but it's pretty silly to be "anti–statistical analysis" and "anti–machine learning" and "anti–automated inference." Long after the bubble is gone, many of these tools will recede to the status of boring utilities, nurtured by open-source weirdos and a few ambitious startups.

That's not the problem. The problem is that all of today's nonsense applications of AI will do lasting damage, separate from these perfectly useful utilities.

There's the obvious, world-threatening harm done by the gigatons of CO_2 emitted by AI data centers to perform the redundant and wasteful training processes and to serve queries.

Then there's the world of labor. We'll see lots of people fired—not because an AI can do their job, but because an AI salesman can convince their credulous bosses to fire them and replace them with an AI that *can't* do their job.

That'll be bad enough. Not only will good people lose their jobs, but the rest of us will lose the value those people produced when they did their jobs. We'll get bad chatbot advice, bad chatbot diagnoses, bad chatbot recommendations.

And then, when the bubble pops, we won't even have *that*. Companies will go bust. The data centers where these giant money- and power-sucking foundation models are housed will go dark. This will be *even* worse than having shitty AI doing important work badly—it will be a time in which that important work won't be done *at all*.

The job of a good AI critic is to help pop the bubble as quickly as possible, before the walls of all our institutions are filled with this digital asbestos that we'll be digging out for generations. To be a good AI critic is to understand the material origins of the bubble, and to strike at the material factors that keep it inflated.

I do a lot of things, but above all, I'm a science fiction writer. I've been selling SF stories since I was a teenager. It's really fun to think about AI, to carry on outlandish thought experiments that challenge us to examine what we think of as intelligence, as agency, as morality. Imaginative exercises, whether undertaken as fiction or as computer code, open your mind to strange possibilities and exciting ideas.

But thought experiments aren't plans. They're not predictions. The fact that you've seen movies and read books with superintelligent AI in them doesn't mean that AI is real. It doesn't mean that it's likely. It doesn't even mean that it's possible.

Many technologists have been inspired by science fiction to create interesting and useful things. The AI scientists who dreamed of better interfaces and came up with the techniques underpinning image generators and large language models did something *incredible*. We will doubtless find ways to refine and use their inventions for many years to come.

But they haven't invented an intelligent being. They haven't set in motion the tools to conjure up a new god or demon. They haven't even invented a tool that can do your job for you.

Reading science fiction to find out what inventions we can expect to see in the future is absurd. It elevates people like me to the status of prophet. Trust me, we do *not* know what the future holds any more than you do.

But there is one thing SF writers know that the AI bosses who are gunning for generational, dynastic wealth by transforming us all into AI-lashed reverse centaurs do *not* know, and that they absolutely don't want *you* to know.

That thing is this: the future is up for grabs. It is not inevitable. AI isn't a genie that can't be put back into a bottle. How we use AI is up to us. Whether we use AI is up to us. The future can be ours, if we never stop remembering that the most important fact about a technology isn't what it does, it's who it does it for, and who it does it to.

Acknowledgments

••••

I write when I'm anxious. The Trump years have been (regrettably) very good for my productivity. But as soothing as writing is for a born compartmentalizer like me, there comes a point when I have to stop working and confront reality, and it is in those times that I rely on the people around me. This book couldn't have been written without the love and care of my wife, Alice, our daughter, Poesy, and my parents, Roz and Gordon Doctorow (my mom is also the best proofreader I know and took a couple runs at this book, finding typos that eluded even the brilliant and talented copyeditors at MCD).

Speaking of MCD, this book is immeasurably better because of the input of my editor, Sean McDonald. This is the second book I've worked on with Sean (the first was *Enshittification*), and his shrewd editorial input proves that the smart ideas he had for *Enshittification* weren't a one-off.

Sean isn't the only talented person at MCD, either. You wouldn't be reading this book but for the incredible success of *Enshittification*, which is largely due to the brilliance of my publicist at MCD, Steve Weil. Steve shares credit with my other publicists around the world, especially Jamie Broadhurst at Raincoast in Canada. Jamie's instincts for how to promote my books in Canada aren't just impeccable, they're *preternatural*, and if you're one of the millions of Canadians who encountered my work, you have Jamie to thank. In the UK, Sam Kelly did a superb job of getting my work in front of every kind of British audience, whisking me from a finance industry luncheon to a dirtbag-left live taping and thence to a Welsh arts variety night.

This book's kick-ass cover is the brainchild of Violet Dine, who was incredibly patient through many rounds of striking the right balance between "whimsy" and "rupture" (thank you to Rodrigo Corral for shepherding us through the process).

The deal for this book was negotiated by superagent Danny Baror, who stepped in while fellow superagent Russ Galen (who has repped me for decades) was sick and did *very* right by me.

Thank you, Danny.

Intellectually, this book owes a great debt to many tech critics (including many who will sharply disagree with some of my conclusions): Ed Zitron, Riley Quinn, Timnit Gebru, Molly Crabapple, Meredith Whittaker, Bruce Schneier, Ed Ongweso Jr, and many, many others. I am also very indebted—as always—to my Electronic Frontier

Foundation colleagues, who have more clarity on technical issues than anyone I know. I'm especially indebted to Kit Walsh for her copyright analysis and Katharine Trendacosta for her labor analysis.

I developed this book in public, on my *Pluralistic* blog, which lives on servers run by Ken Snider, the best sysadmin I've ever worked with. I am not fit to adjust his cronjobs, and I literally couldn't do any of this without him. I am also indebted to Loren Kohnfelder for the Python scripts that help me publish each edition. He sometimes writes to me to point out that I keep thanking him for something he did years ago, but that's rather the point: a technologist can provide you with something that makes your life better *every day*, and for *years and years*.

And finally, thank you to the booksellers and librarians all over the world who have put my books into the hands of readers. As a recovering bookseller and library worker, I salute you. You are the secret legislators of the world.